Study Guide to accompany

INTRODUCTION TO
Chemical Principles
Sixth Edition

— *and* —

BASIC
Chemical Principles
Second Edition

Robert C. Kowerski
College of San Mateo

Edward I. Peters
West Valley College

Saunders Golden Sunburst Series

SAUNDERS COLLEGE PUBLISHING
Harcourt Brace College Publishers

Fort Worth   Philadelphia   San Diego   New York
Orlando   Austin   San Antonio   Toronto
Montreal   London   Sydney   Tokyo

Copyright © 1994, 1990, 1986, 1982 by
Saunders College Publishing

All rights reserved. No part of this publication may be reproduced or transmitted in any form or by any means, electronic or mechanical, including photocopy, recording, or any information storage and retrieval system, without permission in writing from the publisher.

Requests for permission to make copies of any part of the work should be mailed to: Permissions Department, Harcourt Brace & Company, 6277 Sea Harbor Drive, Orlando, Florida 32887-6777.

Printed in the United States of America.

Kowerski/Peters: Study Guide to accompany INTRODUCTION TO CHEMICAL PRINCIPLES, 6/E: BASIC CHEMICAL PRINCIPLES, 2/E

ISBN 0-03-096813-5

7 8 9 0 1 2 3 4 5 6   129   13 12 11 10 9 8 7 6 5

Contents

Chapter 1	Learning Chemistry—NOW!	1
Chapter 2	Matter and Energy	8
Chapter 3	Measurement and Calculations	16
Chapter 4	Introduction to Gases	32
Chapter 5	Atomic Theory and the Periodic Table: The Beginning	38
Chapter 6	The Language of Chemistry	45
Chapter 7	Chemical Formula Problems	52
Chapter 8	Chemical Reactions and Equations	62
Chapter 9	Quantity Relationships in Chemical Reactions	69
Chapter 10	Atomic Theory and the Periodic Table: A Modern View	81
Chapter 11	Chemical Bonding: The Formation of Molecules and Ionic Compounds	91
Chapter 12	The Structure and Shape of Molecules	95
Chapter 13	The Ideal Gas Law and Gas Stoichiometry	103
Chapter 14	Liquids and Solids	112
Chapter 15	Solutions	123
Chapter 16	Reactions that Occur in Water Solutions: Net Ionic Equations	140
Chapter 17	Acid-Base (Proton Transfer) Reactions	148
Chapter 18	Oxidation-Reduction (Electron Transfer) Reactions	157
Chapter 19	Chemical Equilibrium	166
Chapter 20	Nuclear Chemistry	177
Chapter 21	Organic Chemistry	184
Chapter 22	Biochemistry	196
	Answers to Sample Test Questions	205

CHAPTER

1

Learning Chemistry—NOW!

Have you seen *Jurassic Park*?

In the summer of '93 dinosaurs were everywhere. There were dinosaurs in books, in the movies, on commercial television, on PBS pledge drives. There was even a purple dinosaur.

Total spending on dinosaur research during 1992 was about one million dollars. Total United States spending on chemical research and development in 1992 was 161 *billion* dollars.

So where are the chemistry movies and TV shows?

Research and development funding goes to the fields where the benefits of the research can be reaped quickly. The benefits to modern society of chemical research are so many we take them for granted (which is why we don't see many chemistry movies.)

However, without chemical research you couldn't watch a movie or a television show. The film used in the movies was designed by chemists, as were the color phosphors in the television picture tube. You also need electricity to watch a movie or a television show. Chemists monitor the fuel burnt by utility companies to generate electricity to make sure the fuel burns cleanly, to meet air pollution standards.

From the time you wake up until you go back to sleep at day's end, your life *has* been affected by chemistry. We hope in this course to give you some idea of how your life and chemistry interact everyday.

At the end of some chapters in the textbook there are sections called "Everyday Chemistry." These sections give you examples showing how chemistry is a part of your life everyday, and that chemistry enhances your life in society. These examples hold true no matter what your college major.

But the learning comes first. What about this chemistry course? Are you worried? Let's take a look at some of the reasons students get worried, and then we'll show you ways to overcome those worries. We'll list each widespread worry then offer suggestions to help master it.

It's chemistry! What is chemistry, anyway? What do chemists do? In general, chemistry is the study of matter, the world around us. Chemists and chemistry students watch the physical world and ask *"Why is that so?"* As human beings we are innately curious creatures, so asking questions about our surroundings is simply a natural thing to do.

I've forgotten all my math! If you've taken an algebra course, you've already seen all the math you'll need. The text contains a complete math review, including correct calculator usage.

I haven't been back to school in @#?%& years! That's an asset, not a problem at all, because you're now more mature and consequently more educable than you've ever been before. Maturity is an asset in school if you choose to use it. Don't sell yourself short; you're smarter than you've ever been before.

I never learned how to study! Here's the big one. We saved the best for last. The reply to this worry is the remainder of this chapter, which shows you how to *learn*, not just how to *study*.

Learning and studying are two different things. Learning is getting knowledge and understanding of a subject or developing a skill in performing some task (like riding a bicycle.) Studying is one way to learn, but certainly not the only way. Learning is the result of study and other activities. In school what you learn is the bottom line, the only thing that counts.

Unfortunately, much study, much instruction and much experience produce little or no learning. However, there are things you can do to make sure your study effort in chemistry pays off in genuine learning. Pay particular attention if you work in addition to attending school. Let's begin with a few generalizations that apply to *all* subjects, not just chemistry.

Learning Efficiency

If you have homework that requires three hours of genuine learning, how many hours will you have to study to accomplish that task? It will be more than three hours; for some of us, a lot more. How much more depends on your Learning Efficiency (LE).

Learning efficiency is the ratio of minutes of learning to minutes of study time, multiplied by 100. If you get 48 minutes of learning in one hour of study, the learning efficiency is

$$LE = \frac{\text{minutes of learning}}{\text{minutes of study}} \times 100 = \frac{48}{60} \times 100 = 80\% \text{ efficient}$$

The object is to make the numerator as large as possible—maximize learning—while making the denominator as small as possible—minimize the time spent studying. Let's now look at ways to accomplish this crucial task.

Your Study Area

The ideal study area is a quiet place, free from distractions of sight or sound and fully equipped with everything you need, right where you need it. There is nothing to listen to or look at that interrupts concentration, and no time is lost searching for things that are needed.

Let's see how the learning efficiency of Joe College, a typical student, is affected by things known to increase study time without increasing learning. Joe has homework assignments that require three hours of genuine learning time. He figures his learning efficiency is 75%, so he sets aside four hours for the task. That seems logical, but let's look at it closely.

Joe's work area isn't very neat, nor is it well organized. During the four hours Joe spends at least 15 minutes looking for a pencil, paper, his calculator, an assignment sheet, a lecture handout and a badly crumpled scrap of paper that contains the items to be covered on tomorrow's math quiz.

Joe lives in a student dorm. It's not very quiet there; the noise makes it hard to concentrate. Sometimes Joe even leaves his work to make a little noise himself. Noise costs Joe 20 minutes over the four hour period. (If you live at home, the noise is probably different, but it's there in some form.)

Bill stops by for a visit. It lasts 15 minutes.

Joe still smokes. He runs out of cigarettes, and that means a walk to the vending machine. Maybe he can save some time by bumming some from Frank next door. Ten minutes either way. Add five more minutes looking for matches and cleaning up after knocking over the ashtray.

Joe looks at Barbara's picture. She's beautiful. Ten minutes.

"Hey Joe! Telephone." Ten minutes. Joe calls Barbara. Thirty minutes. Total, 40.

Snack time!! Ten minutes, minimum.

Back to noise: There is an insidious kind of noise that commonly affects students. Joe listens to his "personal stereo" as he studies. Joe uses headphones, so the sound can't affect others, but does it affect Joe?

Joe couldn't find a tape he liked (anything by Pearl Jam), so he listened to the radio. Loud. During each of his four study hours, Joe accepted into his thinking about 40% of the ten minutes of commercials (4 minutes), half of the eight minutes of DJ chatter (4 minutes), and one of the two minutes of news. Also, for about 11% of the 38 minutes of music (4 minutes), Joe was gyrating, mouthing words or actually singing, or beating on his drum (desk) with his drumsticks (two pencils.) For one minute, Joe played air guitar. Total cost, 14 minutes. In four hours of study, that comes to 56 minutes. Let's round that off to 40. For 16 of those would-be lost minutes, Joe was visiting, snacking, getting cigarettes, or talking on the telephone.

The four hour study period is now over. Has Joe finished his homework?

Distraction	Cost in Minutes
disorganized work area	15
noise	20
visitor	15
cigarettes	15
Barbara (and other daydreams)	10
telephone	40
snack (frustration food)	10
personal stereo	40
Total	165

Wow! That's 165 minutes lost to distractions in a four hour (240 minute) study period. Only about 75 minutes of learning were salvaged. Joe's learning efficiency is:

$$LE = \frac{\text{minutes of learning}}{\text{minutes of study}} \times 100 = \frac{75}{240} \times 100 = 31\% \text{ efficient}$$

Joe has a problem. His homework assignments require 180 learning minutes, but in four hours of "study" he has finished only 75 of those minutes. That leaves 105 minutes to go. It's nearly midnight, and Joe is tired. If Joe quits now, his homework is less than half done. If he continues to study at the same learning efficiency, he still has almost 6 hours of work ahead. He'll probably fall asleep before then, even given massive amounts of industrial strength coffee. What is poor Joe to do?

Joe's situation is unfortunately not the worst. Crazy Carl across the hall has a test tomorrow. Carl is working 30 hours a week and taking 15 units. Carl is always tired. He hasn't gone to all the lectures, but somewhere he's got copies of a friend's notes. Carl has a brand-new highlighter pen, and he's ready to use it. After highlighting everything, Carl will "do" all the problems that have answers in the back of the book. He will look at the problem, look at the solution in the back, and convince himself that "I can do that!" Finally, Carl tries to study in front of the TV. He looks for either The Brady Bunch (Carl knows *all* the words to the theme....) or a Ren & Stimpy marathon. CC's learning efficiency is about 3%. CC won't be back next term. Maybe Joe won't either.

Nobody's like Carl, and surely you are not Joe. Even so, isn't there a little bit of Joe in all of us? Have you any suggestions for that little bit?

Fatigue and How to Minimize It

Even if you remove from your study area all the distractions that surround Joe College, you still must overcome fatigue. After long hours at a task, people become physically and mentally tired. You will not be physically tired if you get enough sleep (much easier said than done!). If your learning efficiency is high, you will have plenty of time for sleep. High learning efficiency and adequate sleep support work together to help you earn better grades.

Mental fatigue is another matter. After lengthy work periods at the same or similar tasks, you lose sharpness and enthusiasm. You must work harder and longer for a given amount of learning. You cannot avoid this type of fatigue completely, but you can minimize its effects.

Try these ideas to minimize fatigue:

1. If you have several subjects to study, tackle first the most difficult or the least interesting. When fatigue later starts to appear, you'll be doing something else more interesting, and less fatiguing.

2. If you have several subjects to study, and if they are equal in interest and difficulty, rotate them, if it can be done without losing continuity. When you feel yourself losing interest in the first subject, switch to the second subject. Come back to the first subject or go on to the next subject when you tire of the second subject.

3. Take breaks. Study for about 50 minutes, and then take 10 minutes off. (Lectures are 50 minutes because learning drops off rapidly after an hour.) Stretch. Walk around. Do jumping jacks. Snack. Do something else, but watch the time, so you are back in time to start the second hour refreshed at full learning efficiency. Repeat hourly.

4. Work in shorter sessions. You will experience less fatigue in two two-hour study sessions than in one four-hour period. Try a two-hour session in the afternoon and another two-hour period in the evening. When the work is done, you can relax with a clear conscience.

Learning from the Textbook

Your textbook, *Basic Chemical Principles* (BCP) or *Introduction to Chemical Principles* (ICP), and this study guide contain proven learning aids. Take time now to become acquainted with them. You will then be ready to use both books with maximum efficiency.

Preview of Material to Be Learned A textbook assignment should begin with a quick scanning of the material to pick out the highpoints, so you can develop a preliminary idea of its scope and purpose. Your textbook and this study guide (SG) make this crucial task quite easy. Each assignment in this study guide covers the material in about one class lecture. Begin by reading the introductory paragraphs to each assignment in this guide. These paragraphs close with a list of the big new ideas that are introduced in the assignment, exactly the preview that is needed.

Because chemistry courses are highly sequential, the textbook has FLASHBACKS. The Flashbacks are marked with a shining light bulb to remind you of any concept presented earlier that will be required in the new assignment. The section number references from the text are also given in case you wish to review the topic. We strongly recommend such a review whenever necessary.

The specific learning tasks you must perform in each assignment are given by the *Performance Goals*. The Performance Goals (PG) appear throughout the text and are listed by section at the end of each chapter through Chapter 16. Read them carefully. The Performance Goals are the summary highpoints of each chapter. Check page 8 in the textbook to take full advantage of this extremely helpful learning device.

Notetaking Now that you have an idea of what your assignment is about, you are ready to learn. *Learn now, not later*. To help you, the text has Learn

It Now suggestions printed in eye-catching blue. Take the hint and do the learning suggested. As you approach each section that has a performance goal, read it carefully and fix in your thought what to look for as you study. The text has many *Quick Check* questions, set off by a ✔. These force you to think about concepts, and give you instant feedback as to whether or not you really understand them. Summarize the main ideas and write them into your notebook in your own words. If what you see with your eyes stops over in your mind long enough to be analyzed, revised and summarized, you are learning it at that time. Continue through the entire assignment this way. When you finish, you will have a compact set of notes covering the main ideas which you have already learned. When test time comes, you will be able to review the main ideas. That is much easier than learning them for the first time the night before the test.

Unfortunately, most students do not study an assignment in this way. The more common procedure is to sit down with a book and a highlighter pen. Important items are marked, not in condensed form, but in their full textbook presentation. Many pages wind up half colored. You don't have to think about something to realize it is important and highlight it. If you don't think about it, you don't learn it. You have only made yourself a promise to learn it later. When test time comes, you have made yourself so many promises that you just can't keep them all. There is just too much to read and too much to learn in too little time.

Use a highlighter sparingly and intelligently, as a supplement to your handwritten notes. Your notes should have a page reference to the marked material. When you highlight something, you are telling yourself that something is important.

If it's important, learn it. Now!

Summaries and Procedures Look for summaries as you study. In this study guide, you will find a section at the end of each chapter called "Sage Advice and Chapter Clues." This contains a short summary of the material studied, with emphasis to help you avoid the most common pitfalls in the new material. Different types of summaries (set off by a red color) appear throughout the textbook, primarily to help you learn, but also to help you in preparing for a test. These summaries force you to become involved by completing thoughts that summarize the chapter material. Don't forget the list of all performance goals and new terms at the end of each chapter. Many chapters have step-by-step *Procedures* (also denoted by a red color) that help you to do the Example Problems.

Example Problems The only way to learn how to solve chemistry problems is to solve some. To catch your eye, all examples are set off by thin red rules. Check Example 1.1 on text p. 8. The first line of each example states the problem. It's just like a test question; can you solve it? Rule One for solving example problems is the same as Rule One for solving test questions. No peeking! Put a piece of paper over the rest of the example, then go to work at it. When you've got an answer, compare it to the worked-out example under the paper. If you can avoid the temptation to peek (*big* if), you've got the most effective way to learn how to solve chemistry problems.

As you begin learning how to solve chemistry problems, it helps to see clearly that your purpose is **not to solve** the problem, but to **learn how to solve** the problem. You are never finished with an assigned problem until you understand it well enough to solve all other problems of that type. Every assignment in this study guide has a "Skills Quiz" that tests your ability to solve problems, and then gives you detailed solutions and references to the text so you get quick feedback on what areas need more study.

Here are some general hints on solving problems:
1) Be sure you have read and understand the theory or principle behind the problem. Know the definitions of any mathematical relationships you will use, how any equations are written and the units in which they are expressed.

2) As you solve the example problems, be sure you understand each step yourself *before* going on to look at the solution. **This is the time and place to learn how to solve problems.**

3) If you are solving a problem from the end of the chapter, solve the problem *without* referring to an example in the chapter. In particular, do not put one finger at the place of the problem and another finger at the page where a similar example is solved and then flip back and forth, repeating for your problem each step that appears in the example. You will get answers, but no understanding. Instead, if you get stuck, turn from the end-of-chapter problem altogether and work through the matching example from start to finish. When you thoroughly understand the example, close that page of the book, go back to the problem, and solve it completely. If the problem is from the left column at the end of the chapter, compare your solution with the solved problem at the back of the book. If your solution is not correct, find out why.

4) If you have difficulty with a problem in the right column at the end of a chapter, try the corresponding problem in the left column. The principle behind both problems is the same, so the back-of-book answer to the left column problem may help you unravel the right-column problem.

5) Once you get an answer, be sure it is reasonable. (Just because an answer came from a calculator does not make the answer reasonable!) We'll show you how to check your math in Chapter 3.

6) Finally the crucial question: "Did I learn how to solve this problem and others like it?" Even if you have a correct answer, but cannot give a "yes" answer to this question, you are not finished with the problem.

Keep your objective in mind. Your purpose is to learn how to solve problems, not simply to get a correct answer and quickly complete an assignment.

Appendix and Glossary Many students in this course are surprised to learn how much problem solving there is. In addition to Chapter 3 that reviews algebra and problem solving, Appendix I part A is a complete math review, including smart calculator use. Learn well the section on chain calculations with calculators. Students who "know" how to use calculators are often awkward (and incorrect) in these operations.

Instructors often use words that are unfamiliar to students, and then expect students to use the same words intelligently. The Glossary that begins on page G-1 (after the Appendices in the back of the text) is a specialized dictionary that helps to solve the jargon problem. Use both the Glossary, and the list of new "Terms and Concepts" at the end of each chapter regularly.

Learning from the Lecture

What you learn from a lecture depends on what you do before, during, and after the lecture. Let's examine all three.

Before the Lecture Just as a preview of a text reading assignment improves learning from reading the text, so a preview of a lecture improves learning from the lecture. If you know in advance what part of the textbook is to be covered in your next lecture, flip through the pages the night before (or even better, the hour before) the lecture. Glance at section headings and illustrations. Make notes on what you think the main points will be. Try to guess how the ideas go together. Being right or wrong is not important. The act of previewing the lecture prepares you to learn *during* the lecture, rather than after. This takes about ten minutes, but it can save hours of study after the lecture.

During the Lecture What you learn from a lecture depends largely on the quality of the notes you take. In general, the best lecture notes are brief summaries that list the main ideas presented. Phrases are used rather than sentences. Ideally, the notes are in outline form, showing major topics and subtopics. The notes are short, but they include all special conditions that are essential to the main ideas. Good lecture notes also anticipate a follow-up in which the comments are expanded. This is done by writing notes on only one half of the page, or on one of the facing pages in a bound notebook. The remaining space is then available for additional comments.

The first sentence of the last paragraph noted "the notes *you* take." You can't get useful notes from lectures you don't attend. Be there.

To get you into good notetaking habits, the first five chapters of the textbook contain a complete outline, including many hints on how to make a useful outline. Learn these lessons well and they will serve you well in the future, in all your courses.

After the Lecture *This is the crucial time.* H. F. Spitzer, a learning specialist, has demonstrated that a student who waits 24 hours before reviewing lecture notes forgets almost half (46%) of the material presented in the lecture. In two days, 50% is forgotten, and at the end of a week, 62% is gone. By contrast, the student who reviews lecture notes within a few hours of the lecture retains about 98% of what was said, holds 97% a week later, and still remembers more than 90% of the lecture material three weeks (a typical time between exams) after.*

You use the open space in your notebook during the review of the lecture. Write in greater detail the items that were condensed to a few words during the lecture. Check your text for anything you didn't quite understand. Summarize the main points of the lecture. You will find a different color pen very useful when reviewing lecture notes. As in notetaking from the textbook, it is the act of thinking through something to the point that you can write it in your own words that assures learning. Review the lecture just as soon after it is over as possible. Nowhere will you find a better bargain in time and learning. And learning is what it's all about.

Spitzer's work also showed that review two weeks after a lecture increases retention from 20% all the way up to 24%. Moral: You can't cram for exams and hope to do well. (Actually, you *can* cram for quizzes and hour exams and do OK on those. The problem comes when final exams roll around and the knowledge previously gotten so quickly was, alas, lost so quickly. Easy come. Easy go.) You *must* keep up with the course material by reviewing your lecture notes promptly and solving many, many problems. There seems to be no other way. Sigh.

Everyone has a different learning style. An excellent study technique for one person may be unsatisfactory for another. You need not immediately adopt all the suggestions given here, but we do suggest that you consider them. They have worked for other students, and there is every reason to believe most of them will work for you too. Give them a try.

---

* "Studies in Retention", H. F. Spitzer, *Journal of Educational Psychology*

CHAPTER 2

Matter and Energy

Assignment 2A: Describing the Universe

The universe is all mixed up. Few things exist by themselves, so they cannot be studied in isolation and described in precise detail. Separating pure substances from the natural mixtures in which they occur is a major task for a chemist. The results of that separation process must then be described. This requires a special "language" that is used and understood by chemists—and must be learned by chemistry students.

As you approach your first real chemistry lesson, remember to look for the big ideas, and then to organize those ideas into subgroups that are readily remembered, learned and understood. Writing an outline does take some time and practice, but in terms of knowledge gained, that time is very well spent. Also, the performance goals summary at the end of this chapter in the text is a good guide to writing an outline. You should refer to the summaries on text pp. 14 and 21 as we list the big ideas of this assignment:

1) Matter is described by its **properties**, which may be **physical** or **chemical**. Changes in matter may also be physical or chemical.

2) The three common **states of matter** are gases, liquids and solids.

3) Some samples of matter are exactly the same throughout the sample. These are called **homogeneous**. If parts of a sample differ from each other, that sample is **heterogeneous**.

4) If a sample consists of only one kind of matter, it is a **pure substance**. If a sample has two or more kinds of matter, it is a **mixture**.

5) **Pure matter** may be an **element** or a **compound**.

6) One or two letter abbreviations for elements are called **elemental symbols**; elemental symbols are combined to give **chemical formulas**.

Learning Procedures
Study
Sections 2.1–2.2, text pp. 13–22. Focus on Performance Goals 2A–2G as you study.

Answer
Questions and Problems 1–58, text pp. 29–32. Check your answers with those on text pp. A. 13–14.

Take
The Skills Quiz on the next page. Check your answers with those on Study Guide (SG) pp. 12–13.

Assignment 2a Skills Quiz

1) Classify each of the following properties of lithium as physical (P) or chemical (C).

   (a) Melts at 179°C _____

   (b) Has a metallic luster _____

   (c) Reacts quickly with hot water, giving hydrogen gas _____

   (d) Becomes a white powder if exposed to the air _____

   (e) Conducts electricity _____

   (f) Is stable in helium _____

2) Classify the following changes as physical (P) or chemical (C).

   (a) Breaking a dish _____

   (b) Burning coal _____

   (c) Cooking a fish _____

   (d) Digesting an apple _____

   (e) Melting an ice cube _____

   (f) Decaying garbage _____

3) Explain from the standpoint of particle behavior why a Minneapolis resident can store water outside in a strainer in winter, but not in summer.

4) Why can a closed container be half full of a liquid, but never half full of a gas?

5) Classify the following samples of matter as homogeneous (hom) or heterogeneous (het).

   (a) Freshly opened cola _____

   (b) Concrete _____

   (c) Sawdust _____

   (d) The helium in a balloon _____

   (e) Swiss cheese _____

   (f) Poppy seed rye bread _____

6) Carbon monoxide and carbon dioxide are two gases made from carbon and oxygen. Classify each of the following as a pure substance (P) or mixture (M); then classify each pure substance as an element (E) or compound (C).

(a) Carbon _____   (b) Carbon and oxygen _____

(c) Oxygen _____   (d) Carbon and carbon dioxide _____

(e) Carbon dioxide _____ (f) Carbon dioxide and carbon monoxide _____

(g) Carbon monoxide _____

7) How are the physical properties of compounds related to the physical properties of the elements making up the compounds?

8) The density of a liquid was determined at its freezing point. Some of the liquid was frozen, and the density of the remaining liquid was again determined. The second density was greater than the first. Was the original liquid a pure substance or a mixture? Explain your conclusion.

Assignment 2B: Some Fundamental Laws

Matter is anything that has both mass and volume; energy is anything that changes matter. In this assignment you'll meet some basic laws concerning matter, energy and physical and chemical changes.

Here are the big ideas of this assignment:

1) Although **matter and energy** may change in form, their **total amounts** in a physical or chemical change appear to be **constant**.

2) Matter can have two kinds of electric charge, which results in objects being attracted to or repelled by each other. Opposite charges attract; like charges repel.

3) When matter undergoes physical or chemical change, there is also an energy change. An **exothermic change releases energy** to the surroundings; an **endothermic change absorbs energy** from the surroundings.

4) All chemical changes are described by **chemical equations** that always have the arrangement **reactants → products**.

5) Samples of matter may possess **energy of motion (kinetic)** or **energy of position (potential)**.

Learning Procedures

Study

Sections 2.3–2.5, text pp. 22–26. Focus on Performance Goals 2H–2M as you study.

Answer

Questions and Problems 59–78, text pp. 32–33. Check your answers with those on text p. A.14.

Take

the skills quiz below and on the top of the next page. Check your answers with those on Study Guide (SG) p. 13.

Assignment 2B Skills Quiz

1) If a car uses 1/2 gallon (3.1 lb.) of gasoline and 44.7 lb. of air to get you to school, what weight of material comes out of the exhaust pipe during the trip?

2)

Six objects are arranged as shown in the sketch at the left. Two objects are positively charged, two have a negative charge and two are neutral. Draw a solid line between every pair of objects that are attracted to each other, and a dashed line between every pair that repels each other.

3) The chemical equation below is given first using formulas, then using words. Distinguish the products and reactants, and the elemental symbols and formulas of compounds:

$$LiH + H_2O \rightarrow LiOH + H_2$$

lithium hydride and water react to form lithium hydroxide and hydrogen.

4) State whether the following changes are exothermic (exo) or endothermic (endo).

(a) Ice melting _____   (b) Food cooking _____

(c) Gasoline burning _____   (d) Water freezing _____

5) A ball is thrown into the air. Compare the relative amounts of potential energy (PE) and kinetic energy (KE) the ball has at the points:

   (a) as it leaves the thrower's hand

   (b) at the very top of its flight

   (c) halfway down

6) State the Law of Conservation of Energy for a non-nuclear chemical or physical change.

Answers to Chapter 2 Skills Quiz Questions

Assignment 2A

1) Physical (P) properties are a, b and e; chemical (C) properties are c, d and f. Read text pp. 13–15 and review Quick Check 2.1.

2) Physical (P) changes are a, e; chemical (C) changes are b, c, d and f. Physical and chemical properties and physical and chemical changes are best described together so the differences are clearly seen. Read text pp. 13–15, focusing attention on Table 2.1, text p. 14.

3) As solid ice, the particles remain in fixed positions relative to each other, and cannot rearrange themselves so they can fit and flow through the openings in the strainer. As a liquid, which the particles would be in the summer temperatures, this flowing is possible. See text pp. 15–16.

4) Liquids have definite volumes and fill their containers only to the extent of that volume. Gases completely fill any container, regardless of volume. Figure 2.1, text p. 16 illustrates this difference, as noted in the Summary and Quick Check 2.2, on the same page.

5) The only homogeneous substance is helium, choice d. The others have visibly distinct phases. See p. 17 in the text, and Quick Check 2.3.

6) Items a, c, e, g are pure substances (P); items b, d, f are mixtures (M). Items a and c are elements (E); items e and g are compounds (C). Item b is a mixture of two elements; d is a mixture of an element and a compound; f is a mixture of two compounds. The important point is that a compound is a pure substance, one kind of matter, not two. See the first paragraph, text p. 19 and Figure 2.7 on p. 21. A mixture has two or more kinds of matter, which may be two elements, two compounds, or an element and a compound.

7) There is no relationship between the properties of a compound and the properties of the elements in that compound. Read the third paragraph on text p. 21.

8) Physical properties of pure substances are constant. The density of this liquid was not constant. The liquid must therefore have been a mixture. See text pp. 17–18, another illustration of the same idea in Figure 2.3, text p. 17 and review Quick Check 2.4 on text p. 18.

Assignment 2B

1) The total weight of the exhaust is 47.8 pounds. This illustrates the Law of Conservation of Mass, text p. 24, which states that the mass of the reactants in a chemical change equals the mass of the products.

2) 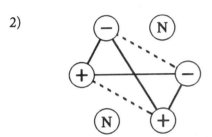 Objects with unlike charges attract each other and objects having the same charge repel. Neutral objects experience neither attraction nor repulsion between each other or with charged objects. See Figure 2.8, text p. 25, and accompanying description.

3) The reactants are lithium hydride, LiH, and water, H_2O; the products are lithium hydroxide, LiOH, and hydrogen gas, H_2. The only element is the H_2 gas; the rest are compounds.

4) The exothermic changes are c and d; the endothermic changes are a and b. You must heat—put energy into—ice to melt it or into food to cook it. There is no question that burning gasoline gives off heat. The release of heat when water freezes is not very apparent, but it is the opposite of ice melting. The difference between exothermic and endothermic changes is on text pp. 23–24; you might also want to review Quick Check 2.6, text p. 24.

5) As it leaves the thrower's hand (a), the ball has low PE, high KE. At the very top of its flight (b), the ball has high PE, low KE. Halfway down, the PE and KE are about equal. Read about the book and the table, text p. 24.

6) In any ordinary change, energy is neither created nor destroyed. Read text pp. 25–26.

Sage Advice and Chapter Clues

In reviewing this and all coming chapters, look for the big ideas, the major concepts around which the chapter is built. If you haven't taken the advice in Chapter 1 and this chapter and made an outline, go back to the start of this chapter and do that now! If your text is covered with highlighter, get thee to thy notebook and make that outline! Identify the big ideas properly, and your study will be organized. Organized study makes it a lot easier for you to understand the principles involved. Only then can you reason from those principles, and use them to solve problems. To help you, the main ideas are listed at the beginning of each assignment, with **boldface terms** to help focus your study.

Chapter 2 acquaints you with words and laws you will use in your study of chemistry. These words and laws are a part of the language of chemistry. As you talk, think or write chemistry, try to use the words and laws correctly. They will soon come freely and accurately, without effort. At the end of each chapter in the text is a section entitled "Key Words and Matching Sets;" match up the words with their definitions, and you've taken a big step in learning the new words in each chapter.

Perhaps the most common mistake with the words in this chapter is to think of a compound as a mixture. Compounds are pure substances, each with its own peculiar physical and chemical properties that never change. To separate a compound into the elements from which it was made requires a chemical change in which the compound is destroyed. A physical change separates a mixture into the pure substances it contains, but the pure substances remain unchanged. Physical changes do not destroy chemical compounds; only chemical changes destroy compounds.

At this time, you may feel anxious about distinguishing between a compound and an element by the name alone. These concerns will be cleared up in Chapter 6. For now, agree that the names of all elements, such as oxygen, iron and sulfur are single words. The names of a few compounds, such as water and ammonia, are also single words. However, the names of most compounds are two words, like carbon dioxide, nitric acid, sodium chloride. In this Study Guide we'll use both words and formulas to describe elements and compounds, to get you used to the idea *before* you're tested on it.

Chapter 2 Sample Test

Instructions: Select the *best* answer for each multiple choice question, and circle the letter of that choice.

1) Copper, Cu, has the properties: (1) it does not react with neon; (2) its density is 8.92 grams per cubic centimeter. Identify the *correct* statement among the following:
   (a) Both (1) and (2) are physical properties.
   (b) (1) is a physical property; (2) is a chemical property.
   (c) (1) is a chemical property; (2) is a physical property.
   (d) Both (1) and (2) are chemical properties.

2) All of the following are physical changes *except*:
   (a) defrosting a frozen pizza.
   (b) putting extra garlic on the pizza.
   (c) slicing the now cooked pizza.
   (d) digesting the pizza by yourself.

3) Which state of matter expands to fill the container that holds it?
   (a) solid   (b) liquid   (c) gas

4) Particle movement is most restricted in which state?
   (a) solid   (b) liquid   (c) gas

5) Among the following, identify the *incorrect* classification:
   (a) A flawless diamond is homogeneous.
   (b) Raisin bread is heterogeneous.
   (c) Filtered spring water is homogeneous.
   (d) A bubbly carbonated beverage is homogeneous.

6) Identify the *incorrect* statement among the following:
   (a) Water is a pure substance, but salt water is a mixture.
   (b) Air is a mixture, but oxygen is a pure substance.
   (c) Salt is a pure substance; sugar is a pure substance, but salt and sugar together are a mixture.
   (d) Nitrogen and hydrogen are both pure substances, but ammonia, the compound formed by them, is a mixture.

7) All of the following are compounds *except*:
   (a) carbon dioxide   (b) sodium chloride   (c) hydrogen   (d) water

8) In the chemical equation given below, identify the reactants and the products, then state which are elements and which are compounds:

   $$H_2 + I_2 \rightarrow 2\,HI$$

   hydrogen and iodine react to form hydrogen iodide.

9) In ordinary chemical and physical changes, the mass of the reactants is _____ the mass of the products.
   (a) always less than   (b) always more than   (c) usually less than   (d) the same as

10) The electrostatic force between a positively charged object and a negatively charged object is one of:
    (a) attraction   (b) repulsion   (c) neither attraction nor repulsion

11) As you make a cup of coffee, you first heat the water to brew the coffee, then let the mixture cool so you can drink it. The physical changes undergone by the water are:
    (a) endothermic; exothermic
    (b) endothermic; endothermic
    (c) exothermic; endothermic
    (d) exothermic; exothermic

12) As you catch a ball that was thrown very high, the potential energy of the ball just before the catch is _____; the kinetic energy of the ball just before the catch is _____.
    (a) high; high   (b) high; low   (c) low; high   (d) low; low

13) In a non-nuclear chemical or physical change, the _____ energy is always conserved, although its _____ sometimes changes.
    (a) form of; total   (b) total; form   (c) kinetic; potential energy   (d) potential; kinetic energy

Check your sample test answers with those on SG p. 205.

CHAPTER 3

Measurement and Calculations

Assignment 3A: Exponential Numbers

Remember back in Chapter 1, we said not to worry if you had forgotten all the math you once knew, because we would review it for you? Here comes the first part of that review....

We learned in Chapter 2 that chemists describe the universe, which means we need to use very large and very small numbers. Exponential notation is simply the easiest way to write these numbers. It's easier than you might think, because having ten fingers and ten toes, we're decimal-based creatures.

Don't overlook the importance of this assignment just because your calculator can do exponential numbers. Historically, many math errors ("decimal slippage") in this chapter and the chapters to follow occur because students enter exponential numbers into calculators incorrectly, *and cannot tell they have made a mistake.*

Section 3.2 introduces an important method of learning problem solving skills, by solving example problems. There are two opaque shields in your text. (If they are gone, make one out of cardboard.) Using a shield correctly, as directed on text p. 37, is the best way to learn to solve problems in this course. Using the shield incorrectly is a quick path to trouble. Try to do the examples with the shield in place, so you can't peek at the answers, and you must work the problem yourself. Remember, you must later work the test problems yourself. Look for the big ideas....

1) Any decimal number can be written in **exponential notation**. Exponential notation expresses a number as a coefficient C (between 1 and 10) multiplied by 10 raised to the e power, in general, $C \times 10^e$. When e is larger than 0, 10^e is larger than 1; when e is smaller than 0, 10^e is smaller than 1. (Remember that $10^0 = 1$.)

2) To add or subtract numbers in exponential notation, adjust the coefficients so all powers of 10 are the same, then add or subtract the coefficients. Don't forget to write the exponents, too.

3) To multiply or divide numbers in exponential notation, multiply or divide the coefficients, then add or subtract the exponents.

Learning Procedures
Study
Sections 3.1–3.2, text pp. 36–41. Focus on Performance Goals 3A–3B as you study.

Answer
Questions and Problems 1–12, text pp. 81–82. Check your answers with those on text p. A.15.

Take
the skills quiz on the next page. Check your answers with those on Study Guide (SG) p. 24.

Assignment 3A Skills Quiz

1) Where decimal numbers are given, write exponential numbers; where exponential numbers are given, write their decimal equivalents.

   21,600 1.23×10^{-6}

   7.69×10^4 0.0000937

2) Perform the following operations; leave 3 digits in your answers.

   $1.67 \times 10^{-4} + 7.8 \times 10^{-5}$ = $9.12 \times 10^2 + 1.4 \times 10^1$ =

   $8.67 \times 10^6 - 4.1 \times 10^5$ = $1.49 \times 10^{-8} - 3.4 \times 10^{-9}$ =

3) Perform the following operations; leave 3 digits in your answers.

   $(4.37 \times 10^{-4})(1.24 \times 10^{-7})$ =

   $(1.79 \times 10^{-2})(5.31 \times 10^4)$ =

   $(9.12 \times 10^6)(3.14 \times 10^3)$ =

   $$\frac{(3.47 \times 10^{-3})(1.46 \times 10^5)}{(8.76 \times 10^{-9})} =$$ $$\frac{6.13 \times 10^7}{(246)(4.71 \times 10^4)} =$$

Assignment 3B: Dimensional Analysis

Dimensional analysis is the easiest way to solve most problems in this chemistry course, and in those that follow. Your text uses dimensional analysis whenever possible. The only mathematical operations required are multiplication and division. When to multiply and when to divide are determined by logical thinking, and the use of measurement units lets you check the correctness of your setup.

The guiding principle of dimensional analysis is that the dimensions on each side of an equation must match. To solve a problem by dimensional analysis you do the following things:

1) Identify the **wanted quantity**, the desired answer to the problem.

2) Identify the **given quantity**, the amount of something on which the entire problem is based.

3) Determine the **unit path**, the **series of logical steps** by which you move from the given quantity to the wanted quantity, the desired answer.

4) Beginning with the given quantity, set up the problem according to the steps in the unit path, using the conversion factor needed for each step.

5) Check the validity of the setup by canceling units.

6) *After* performing Step 5 above, solve for the answer, multiplying and/or dividing as directed in the setup.

The most important skill in dimensional analysis is arriving at the proper setup of the problem, including units. Use reason, logic, and your mathematical skills to arrive at these setups. Don't find the setups by just juggling units. Try to grasp the logic of the problem.

Learning Procedures

Study

Section 3.3, text pp. 41–47. Focus on Performance Goal 3C as you study.

Answer

Questions and Problems 13–22, text p. 82. Check your answers with those on text p. A.15.

Take

the skills quiz below and on the next page. check your answers with those on SG pp. 24–25.

Assignment 3B Skills Quiz

Instructions: In each problem, identify the "given quantity," then solve the problem using dimensional analysis.

1) A 5.5 fathom ship channel has been dug through a shallow stretch of the Mississippi River. If a fathom is six feet, how deep is the channel, in feet?

2) If you made an even exchange of 26 quarters for dimes, how many dimes would you receive?

3) How many seconds are needed for a race horse to run six furlongs at 30.7 miles per hour if one furlong is 40 rods, there are 5.5 yards in one rod, 3 feet in one yard, and 5280 feet per mile?

4) There are 2 tablespoons in 1 ounce, 8 ounces in 1 cup, 4 cups in 1 quart, and 4 quarts in 1 gallon. How many tablespoonsful are there in 2.31 gallons of water in the bottom of an Olde English Rowboat?

Assignment 3C: Mass, Weight and Metric Units

Assignment 3A got you used to thinking in exponential notation, based on the decimal system. The metric system also uses the decimal system to convert between measurements of mass, length and volume.

1) The **mass** of an object **does not change** in different gravities; the **weight** of that object **does change**.

2) The **metric system of measurement** identifies a unit, and then expresses larger or smaller quantities as multiples or submultiples of 10 times that unit. These multiples or submultiples are known by prefixes that may be applied to any measurement unit.

3) The "official" metric unit of mass is the **kilogram, kg**. In the laboratory, mass measurements are commonly expressed in **grams, g**.

4) The metric unit of length is the **meter, m**.

5) The "official" unit of volume is the cubic meter. The more common unit is the **cubic centimeter, cm³**. **Volumes** of liquids and gases are most often expressed in **liters, L centiliters, cL**, or **milliliters, mL**.

6) The important metric prefixes for this course are *kilo-* (1000), *centi-* (0.01) and *milli-* (0.001).

7) By definition, $1 \text{ mL} \equiv 1 \text{ cm}^3$.

8) Because the metric system is decimal based, conversions between larger and smaller metric units involve simply moving the decimal point.

Learning Procedures
Study
Section 3.4, text pp. 48–54. Focus on Performance Goals 3D–3H as you study.

Answer
Questions and Problems 23–40, text pp. 82–83. Check your answers with those on text p. A.15.

Take
the skills quiz below and on the next page. Check your answers with those on SG p. 25.

Assignment 3C Skills Quiz

1) The gravitational field of the moon is 1/6 as large as the earth's gravitational field. An object with a 12 kg mass on earth has what mass on the moon?

2) Give the metric units of mass, length and volume, along with their abbreviations.

3) How many centimeters are in 79.8 meters?

4) A single grain of sand weighs 0.8 milligrams. Express this in grams.

5) Express 3.95 centimeters in millimeters.

6) What is the number of meters in 60.5 kilometers?

7) Find the number of liters in 786 cubic centimeters, cm^3.

8) How many mpedes are in a cpede?

Assignment 3D: Significant Figures

We use two kinds of numbers: exact and approximate. Exact numbers are correct as given. They contain no error, or uncertainty. Examples of these are counting numbers, as how many times you repeat a laboratory procedure, or defined numbers, as the number of centimeters in a meter.

Scientific data are obtained from experiments. These data come from measurements made by reading a scale on a balance, a thermometer, a graduated cylinder or some other device. Measurements are never exact; they involve both imperfect human judgment and imperfect measuring instruments. Measurements are subject to uncertainty, and the best of measurements can always be improved.

The main ideas in this assignment are:

1) **Significant figures** are used to express the **size of the uncertainty** in measurements and in calculations derived from measurements.

2) The number of significant figures in a measurement is the number of digits that are known accurately plus one digit that is doubtful.

3) If a quantity is properly expressed, the doubtful digit is the last digit shown.

4) To count the number of significant figures in a measurement, start with the first nonzero digit on the left and end with the doubtful digit, the last digit shown on the right.

5) To round off a number to the proper number of significant figures, leave the doubtful digit unchanged if the digit to its right is less than 5. Increase the doubtful digit by one if the digit to its right is 5 or more.

6) In addition and subtraction, round off a sum or difference to the first column that has a doubtful digit.

7) In multiplication and division, round off a product or quotient to the same number of significant figures as the smallest number of significant figures in any factor.

8) In chain calculations (text p. A.3), keep intermediate answers in your calculator and round off only the final answer.

Learning Procedures

Study
Section 3.5 text pp. 54–63. Focus on Performance Goals 3I–3L as you study.

Answer
Questions and Problems 41–50, text p. 83. Check your answers with those on text p. A.15.

Take
the skills quiz below and on the next page. Check your answers with those on SG pp. 25–26.

Assignment 3D Skills Quiz

1) The following numbers represent measurements or results calculated from measurements. State the number of significant figures in each.
   (a) 2.96 ___ (b) 18 ___ (c) 0.0558 ___ (d) 3.9×10^4 ___ (e) 7.640 ___

2) A calculated result shows 21.40965312 on the display of a calculator. Round off this number to the number of significant figures required in each case below:

   two _____   three _____   four _____

   five _____   six _____   seven _____

3) Express the sum at the right to the correct number of significant figures.

       27.14
     1224.8
        5.7567
      348.48
     1606.1767

   Answer: _____

4) Express the following result to the correct number of significant figures:

   $$\frac{239.1 \times 4.80 \times 0.0029}{16.508 \times 4.29} = 0.046996667 = \underline{\qquad\qquad}.$$

Assignment 3E: To and from the Metric System, Temperature, Density

Now that you have become familiar with dimensional analysis, the Metric system and significant figures, you are ready to use these tools in calculations involving measurements.

You will use metric measurements in your study of chemistry. In this assignment only will you be concerned with English units, and then just to make a few conversions between them and the metric units.

In the laboratory, temperature measurements are expressed in Celsius degrees, °C. Because Celsius degrees and the familiar Fahrenheit degrees, °F, do not have the same zero point, you cannot use dimensional analysis to convert from one to the other. You must make this conversion by algebra, using Equation 3.3, text p. 66, or some similar equation. The Kelvin temperature scale also does not share a zero point with the Celsius scale, so conversions between these two scales must also be done algebraically using Equation 3.4, text p. 66. All the other calculations in this course can be done using dimensional analysis.

Density is a useful physical property that is a combination of the base units mass and (length)^3, which is volume. Because density is an intrinsic property—not dependent on the size of the sample—it has many uses. For example, canned peas are graded by density; the higher the density, the older and presumably tougher the pea. The peas are placed into tanks of salt water solutions of different densities, the younger peas float, the older peas sink. Sigh.

We'll also look at the cola cans on text p. 35 in this assignment. Look too for these big ideas:

1) The Fahrenheit and Celsius temperature scales are related by the equation $T°_F - 32 = 1.8\, T°_C$.

2) The Kelvin and Celsius temperature scales are related by the equation $T_K = T°_C + 273.15$. Note that there is no degrees sign before the Kelvin temperature.

3) The defining equation for **density** is:

$$\text{density} \equiv \frac{\text{mass}}{\text{volume}}$$

4) Because the defining equation for density is a "per expression," density problems may be solved by either dimensional analysis or algebra.

Learning Procedures
Study
Sections 3.6–3.8, text pp. 63–76. Focus on Performance Goals 3M–3P as you study.

Answer
Questions and Problems 51–88, text pp. 83–85. check your answers with those on text pp. A.15–16.

Take
the skills quiz below and on the next page. Check your answers with those on SG pp. 26–27.

Assignment 3E Skills Quiz

Instructions: Solve each problem in the space provided. A correct setup, beginning with the given quantity, is required for a correct answer.

1) If 30.5 cm = 1 foot, how many feet are in 4.92 meters?

2) If 1 gallon = 3.785 liters, calculate the number of liters in 94.1 gallons.

3) Calculate the mass in grams of a 1.50 lb loaf of bread, if 1 lb = 16 oz and 28.3 grams = 1 ounce (oz).

4) While fishing, you caught a 1.26 kilogram trout. If 2.20 lb = 1 kg, how many pounds is the trout?

5) The mass of 0.922 cm^3 of pure gold is 17.8 g. What is the density of gold?

6) The non-floating cola can on text p. 35 weighed 388.89 g when full and 15.981 g when empty. The volume of the contents was given on the label as 354 mL. Assume 354 mL is also the can volume and calculate the density of this full cola can.

7) A temperature of 42°C is what temperature in °F?

8) A temperature of -12°F is what temperature in °C?

12) A temperature of 84°C is what temperature in K?

13) A temperature of 27K is what temperature in °C?

Answers to Chapter 3 Skills Quiz Questions

Assignment 3A

1) $21{,}600 = 2.16 \times 10^4$ $1.23 \times 10^{-6} = 0.00000123$ Check text pp. 36–39, and
   $7.69 \times 10^4 = 76{,}900$ $0.0000937 = 9.37 \times 10^{-5}$ Examples 3.1 and 3.2.

2) $1.67 \times 10^{-4} + 7.8 \times 10^{-5} = 2.45 \times 10^{-4}$ $9.12 \times 10^2 + 1.4 \times 10^1 = 9.26 \times 10^2$
   $8.67 \times 10^6 - 4.1 \times 10^5 = 8.26 \times 10^6$ $1.49 \times 10^{-8} - 3.4 \times 10^{-9} = 1.15 \times 10^{-8}$
   If these gave trouble, go back to Example 3.4, text p. 40.

3) $(4.37 \times 10^{-4})(1.24 \times 10^{-7}) = 5.42 \times 10^{-11}$
   $(1.79 \times 10^{-2})(5.31 \times 10^4) = 9.50 \times 10^2$ Study Example 3.3, text p.40.
   $(9.12 \times 10^6)(3.14 \times 10^3) = 2.86 \times 10^{10}$
   $\dfrac{(3.47 \times 10^{-3})(1.46 \times 10^5)}{(8.76 \times 10^{-9})} = 5.78 \times 10^{10}$ $\dfrac{6.13 \times 10^7}{(246)(4.71 \times 10^4)} = 5.29 \times 10^0 = 5.29$

   More trouble? Review Appendix I, Part B especially pp. A.5–A.7 in the text.

Assignment 3B

1) GIVEN: 5.5 fathoms   WANTED: feet   PATH: fathoms → feet
   FACTOR: 6 feet/1 fathom $5.5 \text{ fathoms} \times \dfrac{6 \text{ feet}}{1 \text{ fathom}} = 33 \text{ feet}$

2) GIVEN: 26 quarters   WANTED: dimes   PATH: quarters → dimes
   FACTOR: 5 dimes/2 quarters $26 \text{ quarters} \times \dfrac{5 \text{ dimes}}{2 \text{ quarters}} = 65 \text{ dimes}$

3) GIVEN: 6 furlongs, 30.7 mph   WANTED: seconds
PATH: furlongs → rods → yards → feet → miles; hours → minutes → seconds
FACTORS: 40 rods/furlong, 5.5 yd/rod, 3 ft/yd, 1 mile/5280 ft; 60 min/hr, 60 sec/min

$$6 \text{ furlongs} \times \frac{40 \text{ rods}}{1 \text{ furlong}} \times \frac{5.5 \text{ yd}}{1 \text{ rod}} \times \frac{3 \text{ ft}}{1 \text{ yd}} \times \frac{1 \text{ mile}}{5280 \text{ ft}} \times \frac{1 \text{ hour}}{30.7 \text{ miles}} \times \frac{60 \text{ min}}{1 \text{ hour}} \times \frac{60 \text{ sec}}{1 \text{ min}} = 87.9 \text{ seconds}$$

4) GIVEN: 2.31 gallons   WANTED: tablespoons
PATH: gallons → quarts → cups → ounces → tablespoons
FACTORS: 2 tablespoons/ounce, 8 ounces/cup, 4 cups/quart, 4 quarts/gallon

$$2.31 \text{ gallons} \times \frac{4 \text{ quarts}}{1 \text{ gallon}} \times \frac{4 \text{ cups}}{1 \text{ quart}} \times \frac{8 \text{ ounces}}{1 \text{ cup}} \times \frac{2 \text{ tablespoons}}{1 \text{ ounce}} = 591.4 \to 592 \text{ tablespoons}$$

Review text pp. 41–47, particularly Examples 3.5–3.10.

Assignment 3C

1) The object still has a mass of 12 kg on the moon. Mass is a measure of the quantity of matter in an object, and does not change. See text p. 48.

2) The metric unit of mass is the gram, abbreviated g; the metric unit of length is the meter, abbreviated m; the metric unit of volume is the liter, abbreviated L. Find out about these on text pp. 48–50.

For Problems 3–6, all metric-metric conversions are from Table 3.2, text p. 49.

3) GIVEN: 79.8 m   WANTED: cm   PATH: m → cm   FACTOR: 100 cm/m

$$79.8 \text{ m} \times \frac{100 \text{ cm}}{1 \text{ m}} = 7980 \text{ cm}$$ Check Example 3.11, text p. 51.

4) GIVEN: 0.8 mg   WANTED: g   PATH: mg → g   FACTOR: 1 g/1000 mg

$$0.8 \text{ mg} \times \frac{1 \text{ g}}{1000 \text{ mg}} = 0.0008 \text{ g}$$ Example 3.15, text pp. 53–54 also goes smaller unit → larger unit.

5) GIVEN: 3.95 cm   WANTED: mm   PATH: cm → mm   FACTOR: 10 mm/1 cm

$$3.95 \text{ cm} \times \frac{10 \text{ mm}}{1 \text{ cm}} = 39.5 \text{ mm}$$ Example 3.13, text p. 53.

6) GIVEN: 60.5 km   WANTED: m   PATH: km → m   FACTOR: 1000 m/1 km

$$60.5 \text{ km} \times \frac{1000 \text{ m}}{1 \text{ km}} = 60,500 \text{ m}$$ Example 3.12, text p. 52.

7) There are 1,000 mpedes in 1 pede and 100 cpedes in 1 pede, so there are 10 millipedes in 1 centipede. Arthropod fans may groan, but go anyway to Table 3.2, text p. 49. This is like Problem 27, text p. 82.

Assignment 3D

1 (a) 2.96 _3_   (b) 18 _2_   (c) 0.0558 _3_   (d) 3.9×10^4 _2_   (e) 7.640 _4_

Study text pp. 54–58, top, especially Example 3.16 text pp. 57–58.

2) The calculated result 21.40965312 rounds as given below:
   two ____21____   three ____21.4____   four ____21.41____
   five ____21.410____   six ____21.4097____   seven ____21.40965____

   Review text p. 58, and Example 3.17.

3)
$$\begin{array}{r} 27.1 \\ 1224.8 \\ 5.7\;67 \\ \underline{348.4\;8} \\ 1606.1\;67 \to 1606.2 \end{array}$$

   See text pp. 59–60, and Examples 3.18–3.19.

4) Because the factor 0.0029 in the numerator has two significant figures, the calculated answer 0.046996667 must also have only two significant figures, and so rounds to 0.047. See pp. 60–62 in the text, as well as Example 3.21.

Assignment 3E

All metric-metric and Metric-English conversions are from Tables 3.2 and 3.3, text pp. 49, 64.

1) GIVEN: 4.92 m   WANTED: ft   PATH: m → cm → ft
   FACTORS: 100 cm/m, 1 ft/30.5 cm

   $$4.92 \text{ m} \times \frac{100 \text{ cm}}{1 \text{ m}} \times \frac{1 \text{ foot}}{30.5 \text{ cm}} = 16.1 \text{ ft}$$

   Study text pp. 63–65, paying particular attention to Example 3.25, text p. 65.

2) GIVEN: 94.1 gal   WANTED: L   PATH: gal → L   FACTOR: 3.875 L/gal

   $$94.1 \text{ gal} \times \frac{3.785 \text{ L}}{1 \text{ gal}} = 356 \text{ L}$$

   Check problems 69–70, text p. 84.

3) GIVEN: 1.50 lb   WANTED: g   PATH: lb → oz → g
   FACTORS: 16 oz/lb, 28.3 g/oz

   $$1.50 \text{ lb} \times \frac{16 \text{ oz}}{1 \text{ lb}} \times \frac{28.3 \text{ g}}{1 \text{ oz}} = 681 \text{ g}$$

   See Example 3.23, text p. 64.

4) GIVEN: 1.26 kg   WANTED: lb   PATH: kg → lb   FACTOR: 2.20 lb/kg

   $$1.26 \text{ kg} \times \frac{2.20 \text{ lb}}{1 \text{ kg}} = 2.77 \text{ lb}$$

   Example 3.23, text p. 64 and problem 58, text p. 84.

5) GIVEN: 0.922 cm³, 17.8 g   WANTED: D, density   PATH: g ÷ cm³

   $$17.9 \text{ g} \times \frac{1}{0.922 \text{ cm}^3} = 19.3 \text{ g/cm}^3$$

   OR

   GIVEN: 0.922 cm³, 17.8 g   WANTED: D, density   EQUATION: D = m/V

   $$D = \frac{17.8 \text{ g}}{0.922 \text{ cm}^3} = 19.3 \text{ g/cm}^3$$

   See Examples 3.29–3.30, text pp.69–70.

6) GIVEN: 388.89 g full, 15.981 g empty, 354 ml volume   WANTED: D, density
   PATH: g ÷ cm³

   $$(388.89 - 15.981) \text{g} \times \frac{1}{354 \text{ mL}} = 372.91 \text{ g}/ 354 \text{ mL} = 1.05 \text{ g/mL}$$

   OR *continued at top of next page...*

GIVEN: 388.89 g full, 15.981 g empty, 354 mL volume   WANTED: D, density

EQUATION: $D = m/V$ $\quad D = \dfrac{(388.89 - 15.981)\text{g}}{354 \text{ mL}} = 372.91 \text{ g}/354 \text{ mL} = 1.05 \text{ g/mL}$

See Examples 3.29–3.30, text pp. 69–70.

7) GIVEN: 42°C   WANTED: $T_{°F}$ EQUATION: $T_{°F} = 1.8 T_{°C} + 32$
$T_{°F} = 1.8(42) + 32 = 76 + 32 = +108°F$ Text pp. 65–68 and Example 3.27.

8) GIVEN: -12°F   WANTED: $T_{°C}$ EQUATION: $T_{°F} = 1.8 T_{°C} + 32$
$T_{°C} = (T_{°F_2} - 32)/1.8 = (-12 - 32)/1.8 = -44/1.8 = -24°C$ Text pp. 65–68 and Example 3.26.

9) GIVEN: 84°C   WANTED: T_K EQUATION: $T_K = T_{°C} + 273.15$
$T_K = 84 + 273.15 = 357 \text{ K}$ Study Example 3.28, text p. 68.

10) GIVEN: 27 K   WANTED: $T_{°C}$ EQUATION: $T_K = T_{°C} + 273.15$
$T_{°C} = T_K - 273.15 = 27 - 273.15 = -246°C$ Study Example 3.28, text p. 68.

Sage Advice and Chapter Clues

Let's start with calculators. We assume you have one. You don't need an expensive one, but it should be able to display exponential notation and logarithms. A $20 calculator will get you through this course and the general chemistry courses that follow. If you do have a calculator, do you know where its instruction book is? Go find it. Your calculator can do much more and make your life much easier than it is now, if you let it. (More about this later.) Start with the section on chain calculations; to find this section, look up "chain calculations" or "parentheses" in the instruction book's index. If there are no such sections or if your calculator's instruction book is long gone, study well text pp. A.3–A.4 on chain calculations. The important thing is to leave the intermediate answers in the calculator and write down only the final answer. Your time is valuable; let the calculator remember the intermediates.

This is an important chapter, because measurement is what science and engineering are all about. Let's review the highlights of Chapter 3 to see how the assignments lead smoothly from one to the other.

Exponential notation and the metric system go together, because they are both based on powers of ten. There is a very real human tendency to size our measurement units such that all measurements have values ideally between 1 and 1000 (even better, between 1 and 10.) Exponential notation and the metric system let us do that easily. Getting your calculator into exponential notation is simple, if you know what key turns on the exponents. It's marked EE or EEX or EXP. Look for it, then use it with the +/- key to learn how to enter negative exponents with positive coefficients and vice versa. Do it now.

You may have trouble moving the decimal point in the wrong direction when converting from one metric unit to another. You are not likely to make this mistake if you set up the problem, at least in your mind, using dimensional analysis. Remember that there are more little units in a given measurement than there are big units - more centimeters, for example, than meters. Test all your answers mentally to see that they meet this logical requirement.

Dimensional analysis is a powerful problem solving method that you can use in many fields other than chemistry. (Try it on word problems in algebra class.) You may have some difficulty with dimensional analysis at first. For almost all people this difficulty disappears in a week or so. At the end of a month, they are such confirmed "dimensionalysists" they wonder why anyone would solve a problem using any other method. If the procedure seems awkward, keep working with it for a couple of weeks until you master it; don't judge dimensional analysis until you see its full merits.

Your main difficulty is likely to be identifying the given quantity. You must develop this skill carefully, as you will use it continually. Usually, the given quantity is the only value in the question that is *not* a conversion factor, a "per" quantity. Once you have the given quantity, you select your steps in problem solving so you have something logical and meaningful at the end of each step. You will do this automatically if you think your way through the problem, rather than simply plugging in and canceling units.

Significant figures can be a major problem for some students. The digit that gives the most trouble is zero. (What your test score may be if you don't learn significant figures correctly.) Leading zeros, on the far left of a number, are never significant. Trailing zeros, on the far right of a number may or may not be significant. Trailing zeros to the right of the decimal point are always significant, while trailing zeros to the left of the decimal point are probably not significant. (Review the text p. 57 for a detailed discussion of this problem.) When in doubt, put the number in exponential notation, then count the digits. Be very careful when adding or subtracting in a chain calculation. See text p. 81 on this point.

Using a calculator that spews out digits until the display is full makes the "sig fig" problem worse. Don't regard calculator answers as correct. They are not! They are correct to the proper number of significant figures, and no more. Life gets much easier here if you know where your calculator's instruction book is. Casio and Radio Shack owners: Find out what the commands MODE 7 and MODE 8 do. Sharp and TI owners, also look for the MODE or 2nd MODE key. Hewlett-Packard owners: Learn about the FIX and SCI keys.

When a calculator is set to display a fixed number of decimal places, it will round correctly to that number. However, the calculator in this instance will not always display the correct number of significant figures. You must check your calculator, then adjust its display if needed. In exponential notation (exno), the number of significant figures stays constant (as you set it) and the decimal point "floats."

Significant figures are not hard, and the best way to learn them is to use them continuously, even outside of the laboratory. Use them whenever you make a measurement, even when you alone must approve your work.

After learning to express mass and volume units correctly, we combined these quantities to obtain density. Density problems are easily solved using either algebra or dimensional analysis. The only trick to remember when solving density or volume problems is that 1 mL = 1 cm^3.

If solving area (length)^2 or volume (length)^3 problems is confusing, remember that the *entire* conversion factor, not just the unit, must be raised to the power. Try this experiment. Draw a square 1 dm (10 cm) on a side. Now draw a square 1 cm on a side. Cut out the smaller square, put it on the larger square, and estimate how many small squares you would need to cover the larger square completely. You can see that although 10 cm = 1 dm, 10^2 cm^2 = 100 cm^2 = 1^2 dm^2.

Although there are no Performance Goals in Section 3.9, don't overlook this section. Figure 3.6 is a one page summary of this chapter. Finally (whew!), you get to the end of this course the same way you get to Carnegie Hall... Practice, practice, practice, to quote the title of Section 3.10. Take the hint.

Chapter 3 Sample Test

1) Where decimal numbers are given, write exponential numbers; where exponential numbers are given, write their decimal equivalents.

   413,400 6.91×10^7

   0.00103 1.47×10^{-4}

2) Perform the following operations; leave 3 digits in your answers.

   $4.1 \times 10^{-6} + 1.59 \times 10^{-5} =$ $6.7 \times 10^3 + 2.61 \times 10^4 \;=$

   $7.14 \times 10^3 - 3.9 \times 10^2 \;\;=$ $8.34 \times 10^{-1} - 3.6 \times 10^{-2} =$

3) Perform the following operations; leave 3 digits in your answers.

   $(1.16 \times 10^{-3})(6.32 \times 10^{-11}) =$

   $(4.62 \times 10^{-6})(2.17 \times 10^8) \;\;=$

   $(5.71 \times 10^4)(7.45 \times 10^{14}) \;\;=$

   $\dfrac{(9.76 \times 10^{-7})(8.17 \times 10^3)}{(1.23 \times 10^{-1})} =$ $\dfrac{-4.39 \times 10^4}{(107)(7.11 \times 10^1)} =$

4) The Lagrange points are points in space between the earth and the moon; at these points the gravity of the earth exactly cancels the gravity of the moon. What would be the mass and the weight of a 70 kg person at one of these points? (1 kg = 2.20 lb)

5) How many µphones are in a Mphone?

6) Is it colder out when the outside temperature is 0°C or 0°F?

7) How many significant figures are in the measurement 0.099 grams?

8) Round off 2.6034 kilometers to three significant figures.

9) Express the sum at the right to the correct
number of significant figures.

$$\begin{array}{r} 16.08 \\ 0.043 \\ 121.80 \\ \underline{7.99463} \\ 145.91763 \end{array}$$

Answer: _____

10) Express the following to the right number of significant figures:

$$2.193 \times \frac{5.876}{4.88} \times \frac{0.065}{64.06} = 0.0026793350 = \underline{\hspace{3cm}}$$

For Questions 11–16, a correct setup beginning with the given quantity is required for a correct answer.

11) How many dollars can you earn in a part time job in three months if your hourly wage is $6.45, you average 13.6 hours per week, and there are 4.33 weeks in a month?

12) If 2.54 cm = 1 inch, how many centimeters are in 45.0 inches?

13) How many millimeters are in 40.1 meters?

14) Convert 9.45 kilometers to meters.

15) State the number of quarts in 3440 cm³, if 1 liter = 1.06 quart.

16) If 1 ounce = 28.3 grams, how many ounces are in 439 centigrams?

17) A temperature of 14°F is what temperature in °C?

18) A temperature of 71°C is what temperature in °F?

19) A temperature of 312°C is what temperature in kelvins?

20) The full and freely floating cola can on text p. 35 weighed 377.03 g when full. When empty, the can weighed 15.863 g. The volume of the contents was given on the label as 354 mL. Assume 354 mL is also the volume of the can.
   (a) Calculate the density of the full cola can.

   (b) The density of the non-floating cola can was 1.05 g/mL. Comment on the density of the water in the fish tank.

Check your sample test answers with those on SG pp. 205–206.

CHAPTER 4

Introduction to Gases

Everything you have studied so far has been an introduction to chemistry and to this course. No prior knowledge of science has been needed. We expected you to have certain mathematical skills and we reviewed these skills in Chapter 3. You can also find these skills reviewed in Appendix I of the textbook.

As we move into Chapter 4 the cumulative, or "building," nature of chemistry first appears. We now expect you to know what has gone before; we assume you can do all the things listed in the performance goals of the preceding chapters.

Alas, people forget. We understand that while taking a course, you may not always be able to remember ideas presented earlier. To prod your memory, we introduced the FLASHBACK in Chapter 1, text p. 9. The first "real" FLASHBACK was on text p. 89. Did you miss it? Each FLASHBACK, shown by a bright (idea) light, lists earlier topics you will use again; each topic is identified by a section number so you can find it in the textbook if review is needed.

Another thing you will find handy is the GLOSSARY that begins on p. G.1 in the back of the textbook. The glossary is great for a quick reminder of the precise meaning of a term introduced earlier, but not used for several chapters.

To say the least, students lead hectic lives. The most valuable thing you will have this semester is time. Because your time is so valuable, your study *must* be efficient. The FLASHBACKS and GLOSSARY save you time, so use them. End of lecture.

Assignment 4A: Physical Properties of Gases, Gas Measurements

We understand the structure and character of gases better than we understand the structure and character of solids and liquids. Gases give us the clearest picture of the kinetic molecular theory, on which we base our understanding of the differences between gases, liquids and solids. Kinetic molecular theory is introduced on page 15 in the text, which you may wish to review. Look for the following big ideas:

1) The **ideal gas model** describes the activity of molecules in a sample of a gas. This model has been derived to explain the measurable properties of a gas.

2) The **measurable properties of a gas** are pressure, volume, temperature and quantity.

3) In the laboratory, pressure is usually measured by a manometer. While finding pressure from manometer readings is not included in the performance goals, the procedure is given in the caption to Figure 4.4, text page 91.

4) Pressure may be expressed in the following units: atmospheres, centimeters or millimeters of mercury, torr, pascals or kilopascals. In the English system, pressure is measured in inches of mercury or pounds per square inch.

5) Equation 3.4 is shortened to: **K = °C + 273**.

Learning Procedures

Study
Sections 4.1–4.3, text pp. 88–94. Focus on Performance Goals 4A–4B as you study.

Answer
Questions and Problems 1–30, text pp. 108–110. Check your answers with those on text p. A.16.

Take
the skills quiz below and on the next page. Check your answers with those on Study Guide (SG) p. 35.

Assignment 4A Skills Quiz

1) According to the ideal gas model, what causes pressure?

2) Identify the measurable properties of a gas.

3) Express the pressure 851 mm of mercury in torr, cm of mercury, atmospheres, pounds per square inch and kilopascals.

4) Convert 43°C to K. Convert 293 K to °C.

Assignment 4B: Gas Laws, Direct and Inverse Proportionalities

We noted in Assignment 4A that the structure and character of gases are better understood than the structure and character of liquids and solids. This has always been so. Most of the early discoveries by the pioneers 18th and 19th century pioneers of experimental chemistry were based on observations of gases. Some of these experimental pioneers have been immortalized by the use of their names to identify their discoveries. Three appear in this assignment, the high points of which follow....

1) **Gay-Lussac's Law** states that at constant volume, the pressure of a fixed quantity of a gas is **directly proportional** to the absolute temperature, $P \propto T$.

2) **Charles' Law** states that at constant pressure, the volume of a fixed quantity of a gas is **directly proportional** to the absolute temperature, $V \propto T$.

3) **Boyle's Law** states that at constant temperature, the volume of a fixed quantity of a gas is **inversely proportional** to its pressure, $V \propto 1/P$.

4) For a fixed quantity of gas, $PV/T = k$, a constant. Given initial values of all three variables and the final values of two, the remaining value may be calculated using either "proportional reasoning" or the above equation.

5) Standard temperature and pressure, **STP**, are defined as **273 K, 1 atmosphere** pressure.

Learning Procedures
Study
Sections 4.4–4.8, text pp. 94–106. Focus on Performance Goals 4C–4F as you study.

Answer
Questions and Problems 31–72, text pp. 110–112. Check your answers with those on text pp. A.16–17.

Take
the skills quiz below and on the next page. Check your answers with those on SG p. 36.

Assignment 4B Skills Quiz

1) A gas sample has a pressure of 379 torr at 42°C; what is the pressure of this sample if the temperature is changed to 12°C with no volume change?

2) The volume of a gas is 27.6 liters at 15°C. If the temperature is changed to 54°C, with no change in pressure, what is the new gas volume?

3) The volume of a gas is 13.8 liters at 714 torr. If the pressure is raised to 759 torr with no temperature change what is the new gas volume?

4) A gas occupies 6.45 liters at 63°C and 722 torr. What volume would this gas occupy at 97°C and 614 torr?

5) A gas occupies 1.93 L at -22°C and 1.26 atmospheres. What volume would this gas occupy at STP?

Answers to Chapter 4 Skills Quiz Questions

Assignment 4A

1) According to the ideal gas model, pressure is the result of many gas particles moving and colliding with container walls. Read text pp. 89–90.

2) The measurable properties of a gas are: quantity (n); temperature (T); volume (V) and pressure (P). See Section 4.3, page 90 in the text.

3) 851 mm Hg = 851 torr

$851 \text{ mm Hg} \times \dfrac{1 \text{ cm Hg}}{10 \text{ mm Hg}} = 85.1 \text{ cm Hg}$;

$851 \text{ mm Hg} \times \dfrac{1 \text{ atm}}{760 \text{ mm Hg}} = 1.12 \text{ atm}$;

See text pp. 90–93, especially Example 4.1.

$851 \text{ mm Hg} \times \dfrac{1 \text{ atm}}{760 \text{ mm Hg}} \times \dfrac{14.7 \text{ psi}}{1 \text{ atm}} = 16.5 \text{ psi}$

$851 \text{ mm Hg} \times \dfrac{1 \text{ atm}}{760 \text{ mm Hg}} \times \dfrac{101.3 \text{ kPa}}{1 \text{ atm}} = 113 \text{ kPa}$

4) 43°C + 273 = 316 K; 293 K − 273 = 20°C.   Use Equation 4.3, text p. 93.

Assignment 4B

1)

	Volume	Temperature	Pressure	Amount
Initial Value (1)	constant	42°C = 315 K	379 torr	constant
Final Value (2)	constant	12°C = 285 K	P_2	constant

$$379 \text{ torr} \times \frac{(273 + 12) \text{ K}}{(273 + 42) \text{ K}} = 343 \text{ torr}$$

See text pp. 96–98, especially Example 4.3 on 97–98.

2)

	Volume	Temperature	Pressure	Amount
Initial Value (1)	27.6 L	15°C = 288 K	constant	constant
Final Value (2)	V_2	54°C = 327 K	constant	constant

$$27.6 \text{ L} \times \frac{(273 + 54) \text{ K}}{(273 + 15) \text{ K}} = 31.3 \text{ L}$$

See text pp. 98–100, and also Example 4.4, text pp. 99–100.

3)

	Volume	Temperature	Pressure	Amount
Initial Value (1)	13.8 L	constant	714 torr	constant
Final Value (2)	V_2	constant	759 torr	constant

$$13.8 \text{ L} \times \frac{714 \text{ torr}}{760 \text{ torr}} = 13.0 \text{ L}$$

See text pp. 100–103, particularly Examples 4.5–4.6.

4)

	Volume	Temperature	Pressure	Amount
Initial Value (1)	6.45 L	63°C = 288 K	722 torr	constant
Final Value (2)	V_2	97°C = 370 K	614 torr	constant

$$6.45 \text{ L} \times \frac{(273 + 97) \text{ K}}{(273 + 63) \text{ K}} \times \frac{722 \text{ torr}}{614 \text{ torr}} = 8.35 \text{ L}$$

See Example 4.7, text p. 104; ponder the picture on p. 87.

Sage Advice and Chapter Clues

A common mistake in solving pressure-volume-temperature problems is the inversion of the pressure or temperature correction ratio. Be careful in your thinking so you know which pressure or temperature you are coming from and which pressure or temperature you are going to. Using a "tabular analysis" as shown on text p. 97 helps to organize your thoughts. Some students label each quantity in the problem with a subscript $_1$ or $_2$ after reading the problem. Find a way that works for you.

You will understand these problems better if you reason your way through the correction ratios. If you are unsure that your set-up is correct, compare it with the set-up you obtain by solving the problem algebraically using Equations 4.15 and 4.16 (text p. 104;) the correct set-ups will be identical. If they are not, recheck both until the error is found and corrected.

Oh yes, and remember to change all temperatures from °C to K....

Chapter 4 Sample Test

Instructions: For 1, 2, choose the letter of the *best* choice. Answer the other questions in the space provided.

1) Pick the statement about the ideal gas model that is *incorrect*:
   (a) The volume of the particles, or molecules, is negligible compared to the volume occupied by the gas.
   (b) There are large attractive forces between molecules in an ideal gas.
   (c) Intermolecular collisions occur without loss of kinetic energy.
   (d) Gas molecules are in constant motion.   (e) Gas molecules are independent of each other.

2) A pressure of 0.836 atmospheres is equal to:
   (a) 836 torr   (b) 63.5 cm Hg   (c) 732 mm Hg   (d) 0.846 kPa   (e) none of these

3) A gas occupies 0.610 L at 0.103 atm pressure. What volume will the gas occupy at 1.62 atm pressure, if the temperature is held constant?

4) A gas exerts 462 torr pressure at -1°C. When the pressure of this gas at constant volume is changed to 591 torr, what is the final temperature of the gas?

5) A gas occupies 2.14 L at 40°C; what volume does this gas occupy if the temperature is lowered to 20°C, with pressure remaining constant?

6) Initially a gas occupies 4.80 L at 744 torr and 32°C. What volume will it fill at 811 torr and 64°C?

7) A gas occupies 1.24 L at STP. Find the volume it would occupy at 21°C and 1.21 atmospheres.

Check your sample test answers with those on SG pp. 206–207.

CHAPTER 5

Atomic Theory and the Periodic Table: The Beginning

Assignment 5A: The Nature of Atoms

Can you divide a grain of sand into smaller and smaller grains? Is matter (such as the sand) continuous, or is there an ultimately small grain of matter? Philosophers argued these questions for 2200 years in the absence of experimental evidence. Today we have evidence that atoms do exist, and so everybody knows about atoms. Chemists and chemistry students know a little more about atoms than most people do. To understand chemistry, you must know how atoms take part in chemical changes.

The main ideas in this assignment are:

1) An **atom** is the smallest part of an element.

2) Dalton's atomic theory identified five properties of atoms; we now know that two of those properties were incorrect.

3) The three major parts of an atom are the **proton**, **neutron** and **electron**. Each of these has its own mass and electrical charge.

4) Most of the mass and all of the positive charge in an atom is concentrated in an extremely tiny nucleus. The lightweight electrons occupy a relatively huge space outside of the nucleus.

5) **Isotopes** are atoms of the same element that have different masses. Isotopes are identified by symbols of the form $^A_Z Sy$ or $^A Sy$.

Learning Procedures

Study
Sections 5.1–5.4, text pp. 115–121. Focus on Performance Goals 5A–5F as you study.

Answer
Questions and Problems 1–26, text pp. 131–132. Check your answers with those on text p. A.17.

Take
the skills quiz on the next page. Check your answers with those on Study Guide (SG) p. 41.

Assignment 5A Skills Quiz

1) Describe atoms according to the atomic theory proposed by Dalton. Identify those parts of the theory we now consider to be incorrect.

2) Given the choices: (A) +1 charge   (B) -1 charge   (C) 0 charge
                             (D) about 1 amu   (E) almost 0 amu
match the letters of the choices that best describe each of the following:

neutron _____   electron _____   proton _____

3) The gold foil used in the Rutherford scattering experiment was about 2000 atoms thick. Why did most of the alpha particles pass straight through that solid foil?

4) Finish the following table. You may refer to the Table of Elements, inside back cover of the text.

Nuclear Symbol	Number of Protons	Number of Neutrons	Number of Electrons	Atomic Number	Mass Number
^{34}S					
	23	28			
				38	88
			13		27

Assignment 5B: Atomic Mass, the Periodic Table, Names & Symbols

One of the things scientists do to help them understand nature is to organize their observations and seek out patterns or regularities that may be present. Chemists have been doing this for a century and a half. The resultant understanding of atoms and elements underlies the chemical industry that gives so many benefits to humanity. (Unfortunately, some of this knowledge is abused in ways that are not so beneficial.) One of the most remarkable organizational devices ever developed was the periodic table, which was first arranged according to the atomic masses of the elements. In this assignment, you will study atomic masses, the periodic table, and the names and symbols of some common elements.

1) The **atomic mass unit** (amu) is defined as exactly 1/12 the mass of a single atom of carbon-12.

2) The **average mass of the atoms of an element** as found in nature is called the **atomic mass**. The atomic mass may be calculated from the masses of the natural isotopes of that element and the percentage abundance of each element.

3) The periodic table arranges the elements into periods and groups in order of atomic numbers and periodic recurrence of physical and chemical properties.

4) Each box in the periodic tables in your textbook gives the atomic number (Z), the symbol, and the atomic mass of an element.

5) The names and symbols of 35 common elements are to be learned, using the periodic table as a memory aid.

Learning Procedures

Study
Sections 5.5-5.7, text pp. 121–128. Focus on Performance Goals 5G-5K as you study.

Answer
Questions and Problems 27–52, text pp. 132–133. Check your answers with those on text pp. A.17–18

Take
the skills quiz below and on the next page. Check your answers with those on SG pp. 41–42.

Assignment 5B Skills Quiz

Instructions: You may refer to a periodic table when answering the questions.

1) What is an atomic mass unit?

2) The mass numbers of two naturally occurring isotopes of imaginary element X and their relative abundances are, respectively, 47.0 amu–61.0% and 50.0 amu–39.0%. Calculate the average atomic mass of element X.

3) In which period (_____) and group (_____) is sulfur in the periodic table?

4) For the element for which Z = 81, write the symbol _____; period _____; group_____; average atomic mass _____.

5) Write the chemical symbol for each element given, and the name of the element for each symbol given.

F _____   potassium _____

Cl _____   iron _____

Ag _____   barium _____

Answers to Chapter 5 Skills Quiz Questions

Assignment 5A

1) (a) Elements are made up of tiny, individual particles called atoms. (b) Atoms can neither be created nor destroyed. (c) All atoms of an element are identical in every respect. (d) Atoms of each element are different from atoms of any other element. (e) Atoms of one element can combine with atoms of another element to form compounds. When they do, they usually combine in a ratio of small, whole numbers. We now know that items (b) and (c) are incorrect. See text pp. 115–116.

2) neutron, C, D; electron, B, E; proton, A, D   See Table 5.1, text p. 117.

3) The alpha particles passed through the foil because most of the volume of an atom is empty space. Check text p. 118, and Figure 5.3.

4)

Nuclear Symbol	Number of Protons	Number of Neutrons	Number of Electrons	Atomic Number	Mass Number
^{34}S	16	18	16	16	34
^{51}V	23	28	23	23	51
^{88}Sr	38	50	38	38	88
^{27}Al	13	14	13	13	27

See equation 5.1, text p. 119, and Example 5.1 text p. 120.

Assignment 5B

1) The atomic mass unit is exactly 1/12 of the mass of one carbon-12 atom. See text p. 121.

2) (47.0)(0.610) = 28.7
   (50.0)(0.390) = <u>19.5</u>
   average atomic mass = 48.2 amu   See Examples 5.2, 5.3, text pp. 122–123. Go over Table 5.3 carefully to make sure you have no calculator typing errors.

3) The element sulfur is in period 3, group 6A in the periodic table. Check text pp. 123–126.

4) If Z = 81, symbol is Tl, element is in period 6, group 3A, and has an average atomic mass of 204.37 amu. Review Example 5.5, text p. 126.

5) F, fluorine; Cl, chlorine; Ag, silver; potassium, K; iron, Fe; barium, Ba. If you had trouble with these, study Figure 5.8, text p. 127.

Sage Advice and Chapter Clues

In 1912, Max von Laue found the first evidence that atoms were real. Not until 1982 did the scanning tunneling microscope (See Everyday Chemistry in Chapter 10 of the text.) give us the first "pictures" of atoms. As a result, our understanding of atomic theory is based on abstract ideas, models. A scientific model is a mental system that is like something that can be seen with the eyes, or that can be built. At the very least, a model must be like something we can imagine to be big enough to see or physically handle. It is easier to understand what we cannot see if we can compare it with something we can see.

Once you form a mental model, you return to the laboratory *to test it*. You try to determine what should happen in beakers and test tubes if your model is correct. Using your new experimental observations, you do one of three things with your model: (1) you throw it out; (2) you modify it; or (3) you keep it. You always hope for case (3), but often more is learned if case (1) or (2) apply. The mark of a scientist is not as much the ability to devise a model, but rather being ready and willing to modify or discard it if it fails in the laboratory. Whatever happened to cold fusion, anyway?

We have already seen two parts of Dalton's atomic theory discarded because of experimental facts that contradict his model. However, the three parts that remain are foundation stones of modern chemistry. The nuclear and planetary models also contributed to our present understanding of atoms. In Chapter 10, you will see that the planetary model has been replaced by a newer model that conforms more closely to experimental evidence.

Assignment 5B introduced you to the "chemist's cheat sheet," the periodic table. The periodic table is a marvelous source of chemical information. Later you will learn to take full advantage of the information gathered there. But a word of caution is in order. In this course, and in most others, the form of the periodic table is that which is on the inside front cover of the textbook, or on your opaque shield. Usually, this is the form that will be available to you. Please don't get in the habit of using a store-bought periodic table with additional information crowded onto it in four beautiful colors. Avoid marking your own periodic table, too, so you can learn to find everything you need from a "clean" periodic table. Become familiar with the periodic table. It is a visual summary of lots of information; it can be a big help. (Isn't that a great picture of Dmitri Mendeleev on text p. 123? Mendeleev was a political radical who lectured on politics and otherwise insulted the Czar during a private audience. As a result, Russian chemistry textbooks did not mention Mendeleev until after the 1917 Russian Revolution.)

Students sometimes complain that "Chemistry is a memory course. You gotta memorize everything." Indeed, that last performance goal in Assignment 5B requires you to memorize the names and symbols of some common elements. Is this then, just a memory course, understanding not needed? Read on, please.

How do you know your own name? You were not born with that knowledge, so you must have learned it somewhere.... You memorized it; you heard people call you by that name so often, it became second nature to respond to it.

The same plan works in chemistry. You must start to learn the rules of a name, or a science, somewhere, and that somewhere is with memorization of a relatively few facts. Do that memorizing *now*; don't skimp on the time needed to learn these element names. In a couple of weeks, those element names will be like the names of friends; you'll then use those names as the basis for better understanding of chemistry the rest of the term.

Chapter 5 Sample Test

Instructions: Select the *best* answer for each multiple choice question, and circle the letter of that choice. You may refer to a "clean" periodic table.

1) Dalton's second postulate states that atoms can neither be created nor destroyed. This postulate supports the _____.
   (a) Law of Definite Composition
   (b) Law of Conservation of Mass
   (c) existence of isotopes
   (d) nuclear model of Lord Rutherford

2) Which subatomic particle, p, n or e, is the lightest?
   (a) p
   (b) n
   (c) e

3) If a very small amount of neon gas is placed in a previously empty glass tube, then the tube is sealed, the neon glows when electricity is passed through the tube. While glowing, particles from the neon flow towards *both* the negative and positive ends of the tube. This particle flow shows that....
   (a) neon atoms are indivisible.
   (b) neon is a gas made up of atoms.
   (c) neon atoms are electrically neutral.
   (d) neon atoms contain positively and negatively charged parts.

4) Rutherford's scattering experiments showed that the electrons:
   (a) are outside the nucleus in almost vacant space.
   (b) are packed tightly together in the nucleus.
   (c) are packed tightly together outside of the nucleus.
   (d) occupy very little space in the nucleus.

5) Two atoms are identified by the symbols ^{22}Ne and ^{23}Ne. Which of the following statements about these atoms is *false*?
   (a) The atoms contain the same number of protons.
   (b) The masses of the atoms are different.
   (c) ^{22}Ne has fewer neutrons than ^{23}Ne.
   (d) The number of electrons equals the number of protons in ^{22}Ne, but not in ^{23}Ne.

6) Select the *correct* statement about $^{43}_{22}$Ti.
   (a) The nucleus contains 22 protons and 26 neutrons, mass number = 48.
   (b) The nucleus contains 22 protons and 21 neutrons, atomic number = 22.
   (c) The nucleus contains 21 protons and 22 neutrons, mass number = 43.
   (d) The nucleus contains 22 protons and 43 neutrons, atomic number = 22.

7) An atomic mass unit (amu) is...
   (a) the mass of an atom in grams.
   (b) 1/12 the mass of an atom of carbon-12
   (c) the mass of an atom of carbon-12.
   (d) the mass of an atom compared to the mass of a carbon-12 atom.

8) The mass numbers of two natural isotopes of an imaginary element and their relative abundances are, respectively, 94.0 amu, 82.4% and 99.0 amu, 18.6%. What is the average atomic mass of this imaginary element?
   (a) 94.0 amu   (b) 95.9 amu   (c) 96.5 amu   (d) 98.6 amu

9) Select the *correct* placement in the periodic table for nitrogen, N.
   (a) group 5A (15), period 1
   (b) group 2, period 5A (15)
   (c) group 7, period 2A (2)
   (d) group 5A (15), period 2

10) Select the *correct* statement about an element for which the atomic number is 16.
    (a) The element is sulfur, S; for which A = 32, group 6A (16), period 3.
    (b) The element is sulfur, S; for which Z = 32, group 6A (16), period 3.
    (c) The element is oxygen, O; for which A = 16, group 6A (16), period 2.
    (d) The element is oxygen, O; for which A = 8, group 6A (16), period 2.

11) Write the name of the element for each symbol given, and the chemical symbol for each element given.

Br _____   sodium   _____

Mg _____   nickel   _____

Pb _____   phosphorus _____

Fe _____   calcium   _____

Ag _____   silicon   _____

K _____   fluorine   _____

Hg _____   manganese   _____

Check your sample test answers with those on SG p. 207.

CHAPTER 6

The Language of Chemistry

Chemists are busy, creative people. So far, chemists have discovered 109 elements. From these 109 different types of atoms, chemists have made or discovered over 11,000,000 compounds since 1965. People have developed systems to help them understand how atoms join to form compounds, and how these compounds are named. No one (not even your chemistry instructor) wants to memorize eleven million different names....

By the end of this chapter you should be able to write the names and formulas of hundreds—perhaps thousands—of chemicals. You won't be able to reach this level by trying to memorize all those names and formulas. Instead, memorize the relatively few rules of chemical nomenclature, and then concentrate on learning how to apply those rules. If you know the system, you can figure out a name or formula when given the other. If you try to memorize names and formulas without learning the system, you will never know more than what you once memorized—and have not forgotten.

This is the first (but not the last) time we'll give the summary for this chapter. Take it to heart:
Learn the System!

Assignment 6A: Reviewing Elements, Naming Compounds Made from Two Nonmetals

Back in Section 5.7 (text p. 126), you learned the symbols of some common elements in the periodic table. Any chemist can tell you that some elements are trickier than others. For chemistry students, those trickier elements are the seven that exist as diatomic molecules at room temperature. In this assignment you'll learn to write the formulas of those elements correctly. You'll also learn to write the names and formulas of some molecules made from two different atoms.

Here come the big ideas...

1) The names and formulas of the 35 elements in Figure 5.8, text p. 127 should already be in memory. The seven diatomic elements, H_2, N_2, O_2, F_2, Cl_2, Br_2 and I_2 must be learned before it's too late.

2) **Two nonmetals** form chemical bonds with each other to form **binary molecular compounds**. The name of a binary molecular compound is the name of the first element followed by the name of the second element, modified with an *-ide* suffix. Prefixes are used to indicate the number of atoms of each element in the molecule.

3) Two common binary molecular compounds with nonsystematic names are **water, H_2O** and **ammonia, NH_3**.

Learning Procedures

Study
Sections 6.1–6.3, text pp. 137–140. Focus on Performance Goals 6A–6C as you study.

Answer
Questions and Problems 1–10, text p. 165. Check your answers with those on text p. A.18.

Take
the skills quiz below. Check your answers with those on Study Guide p. 49.

Assignment 6A Skills Quiz

Instructions: For each name given, write the formula; for each formula given, write the name.

sodium _____	Cl_2O _____
zinc _____	Pb _____
iodine _____	NI_3 _____
diphosphorus pentoxide _____	Fe _____
ammonia _____	Sn _____
manganese _____	H_2O _____
calcium _____	$SiCl_4$ _____
fluorine _____	S_4N_2 _____
hydrogen _____	O_2 _____
nitrogen _____	F_2 _____
bromine _____	Cl_2 _____

Assignment 6B: Naming Ions

In this assignment you'll learn to name ions. The big ideas follow:

1) Ions are charged particles. A **cation** has a **positive charge**; an **anion** has a **negative charge**.

2) The **oxidation state** of a monatomic substance is the same as the charge on the substance.

3) The name of a monatomic cation is the name of the element, followed by the word "ion." If the element can form more than one monatomic cation, its oxidation state is added to the elemental name. The oxidation state is written in parentheses immediately after the name. For example, Fe^{2+} is iron(II), pronounced "iron two ion."

4) An **acid** ionizes in water to give **H⁺** and an anion. A general equation to describe acid ionization is **HA → H⁺ + A⁻**.

5) **Binary acids** are made from hydrogen and one nonmetal element. Binary acids are named **hydro----ic** acid.

6) The name of a **monatomic anion** is the name of the element changed to end in **-ide**, followed by the word ion.

7) **Oxyacids** contain hydrogen, oxygen, and a third nonmetal atom.

8) The nomenclature system for oxyacids and the ions derived from their ionization includes prefixes and suffixes. Acids that end in -*ic* give anions that end in -*ite*. An acid with an -*ic* ending always has 1 more oxygen atom than the corresponding acid with an -*ous* ending. The system: **-ic → -ate; -ous → -ite; -ic > -ous.**

Learning Procedures

Study
Sections 6.4–6.6, text pp. 140–152. Focus on Performance Goals 6D–6F as you study.

Answer
Questions and Problems 11–26, text pp. 165–166. Check your answers with those on text p. A.18.

Take
the skills quiz below. Check your answers with those on SG p. 49.

Assignment 6B Skills Quiz

Instructions: For each name given, write the formula; for each formula given, write the name.

chromium(II) ion _____

hydrosulfuric acid _____

potassium ion _____

iodous acid _____

nitride ion _____

copper(II) ion _____

selenic acid _____
(from selenium, Z = 34)

hydrogen sulfate ion _____

hypoiodous acid _____

calcium ion _____

nitrous acid _____

silver ion _____

hydroiodic acid _____

Pb^{2+} _____

F^- _____

O^{2-} _____

$HClO_3$ _____

H_2Te _____
(Te is tellurium, Z = 52)

CO_3^{2-} _____

$H_2PO_4^-$ _____

H_2SO_4 _____

$HBrO_4$ _____

Fe^{3+} _____

HNO_3 _____

Sn^{2+} _____

Hg_2^{2+} _____

Assignment 6C: More Names, More Formulas, Ionic Compounds, Too...

You've learned to name cations in Assignment 6A, acids and their anions in Assignment 6B. In this assignment you'll combine cation and anion naming skills to name ionic compounds called salts. Look for these high points:

1) The **ammonium** ion is NH_4^+; the **hydroxide** ion is OH^-.

2) To write formulas for ionic compounds, write the symbol for the cation, then the anion. Use subscripts to take these ions so their total charge is zero.

3) To name ionic compounds, name the cation, then the anion.

4) **Hydrates** are ionic compounds that exist with a **definite number of water molecules** in their crystal structure. Waters of hydration are indicated by the · symbol before the number of waters.

Learning Procedures

Study
Sections 6.7–6.10, text pp. 152–160.

Answer
Questions and Problems 27–78(!), text pp. 166–169. Check your answers with those on text pp A.18–20. Also, spend serious time working Tables 6.12 and 6.13 *completely*.

Take
the skills quiz below. Check your answers with those on SG pp. 49–50.

Assignment 6C Skills Quiz

Instructions: For each name given, write the formula; for each formula given, write the name.

ammonium sulfide	_____	$BaCl_2$	_____
calcium bromide	_____	NaI	_____
potassium oxide	_____	$NH_4C_2H_3O_2$	_____
magnesium phosphate	_____	K_3N	_____
zinc nitrate	_____	$MgSO_3$	_____
iron(II) carbonate	_____	$Ca(HSO_4)_2$	_____
potassium hydrogen phosphate	_____	AlF_3	_____
potassium nitrate	_____	LiOH	_____
beryllium acetate	_____	$FePO_4$	_____
sodium sulfate 10-hydrate	_____	$Cr_2(CO_3)_3$	_____
$Ca(NO_3)_2 \cdot 4\,H_2O$	_____		

Answers to Chapter 6 Skills Quiz Questions

Assignment 6A

sodium, Na Figure 5.8, text p. 127
zinc, Zn Figure 5.8, text p. 127
iodine, I_2, text p. 138
diphosphorus pentoxide, P_2O_5, text p. 127
ammonia, NH_3, text p. 140
manganese, Mn, Fig. 5.8, text p. 127
calcium, Ca, Figure 5.8, text p. 127
fluorine, F_2, text p. 138
hydrogen, H_2, text p. 138
nitrogen, N_2, text p. 138
bromine, Br_2, text p. 138

Cl_2O, dichlorine oxide, text pp. 272-273
Pb, lead, Figure 5.8, text p. 127
NI_3, nitrogen triiodide, text pp. 138–139
Fe, iron, Figure 5.8, text p. 127
Sn, tin, Figure 5.8, text p. 103
H_2O, water, text p. 140
$SiCl_4$, silicon tetrachloride, text p. 252
S_4N_2, tetrasulfur dinitride, text p. 273
O_2, oxygen, text p. 138
F_2, fluorine, text p. 138
Cl_2, chlorine, text p. 138

Assignment 6B

chromium(II) ion, Cr^{2+}, Fig. 6.2, text p. 142
iodous acid, HIO_2, Table 6.3, text p. 149
selenic acid, H_2SeO_4, Example 6.9, text p. 150
hydrogen sulfate ion, HSO_4^-, Table 6.5, text p. 152
hypoiodous acid, HIO, Example 6.7, text p. 148
nitrous acid, HNO_2, Example 6.8, text p. 149
silver ion, Ag^+, Figure 6.2, text p. 142
mercury(I) ion, Hg_2^{2+}, text p. 142
This is a tricky one. Read the text carefully.
hydroiodic acid, HI, text p. 144
HI may also be called hydriodic acid, see text p. 139.

Pb^{2+}, lead(II) ion, text p. 142, see Figure 6.2.
$HClO_3$, chloric acid, Table 6.3, text p. 147
H_2Te, hydrotelluric acid, Example 6.4, text p. 144
CO_3^{2-}, carbonate ion, Table 6.2, text p. 146
$H_2PO_4^-$, dihydrogen phosphate ion, text p. 151
$HBrO_4$, perbromic acid, Example 6.7, text p. 148
Fe^{3+}, iron(III) ion, text p. 142
HNO_3, nitric acid, Table 6.2, text p. 146

Sn^{2+}, tin(II) ion, Figure 6.2, text p. 142

Assignment 6C

ammonium sulfide, $(NH_4)_2S$
calcium chloride, $CaCl_2$
potassium oxide, K_2O
magnesium phosphate, $Mg_3(PO_4)_2$
zinc nitrate, $Zn(NO_3)_2$
iron(II) carbonate, $FeCO_3$
potassium hydrogen phosphate, K_2HPO_4
ammonium sulfide, $(NH_4)_2S$
calcium bromide, $CaBr_2$
potassium nitrate, KNO_3
magnesium phosphate, $Mg_3(PO_4)_2$
sodium sulfate 10-hydrate, $Na_2SO_4 \cdot 10\, H_2O$

$BaCl_2$, barium chloride
NaI, sodium iodide
$NH_4C_2H_3O_2$, ammonium acetate
K_3N, potassium nitride
$CuCl_2$, copper(II) chloride
$MgSO_3$, magnesium sulfite
$Ca(HSO_4)_2$, calcium hydrogen sulfate
$BaCO_3$, barium carbonate
NaI, sodium iodide
AlF_3, aluminum fluoride
LiOH, lithium hydroxide

$Ca(NO_3)_2 \cdot 4\ H_2O$, calcium nitrate 4-hydrate is preferred, but calcium nitrate tetrahydrate or 4-water may also be acceptable. Check with your instructor.

If you missed any names of the ionic compounds in the Assignment 6C skills quiz, read text pp. 157–158, and study Example 6.16, text p. 158. If you missed any formulas, read text pp. 154–157 and review Tables 6.6 and 6.7, text pp. 153, 154. Study Examples 6.11 and 6.12, text pp. 155–156 if you don't understand how to write formulas that need parentheses. Go back and take another look at Table 6.8, text p. 160; there's a *lot* of good information there. The last two hydrate problems are covered in Section 6.10, text p. 159, Example 6.17.

Sage Advice and Chapter Clues

Sometimes to know what to do, you've also got to know what *not* to do. Here's a list of the most frequent student nomenclature errors.

1) Students have not learned the formulas of the diatomic elements. To remember the elements that form diatomic molecules, just remember the sentence: "Horses Need Oats For Clear, Brown Eyes." The first letter or two of each word is the symbol for the elements that form diatomic molecules, in order of increasing atomic number. Just don't write "eyeodine" for iodine, and you'll never miss the diatomics. You'll even get the symbol for fluorine correct; it's F, as in For, not Fl. Please don't groan. It may be desperate doggerel, but it works....

2) In naming binary molecular compounds, the Greek prefixes for number must be used before the name of the element that contributes more than one atom to the molecule. The prefix *mono-* may be used if only one atom of an element is present; in the absence of a prefix one atom is understood. Go back to Table 6.1, text p. 139 to find these prefixes.

3) Students try to memorize all the formulas or names, or even ions. That's a waste of time, if not impossible. *Learn the system* and limit your memorization to the few essential starting points.

4) Some students fail to appreciate the importance of Tables 6.2–2.4, text pages 146–152. These summarize the nomenclature system for acids and anions, on which the names of many ionic compounds are based. These tables should be the focal points of your memorization efforts. (Put another way, if you don't know Table 6.2 in particular, you're history....)

3) Students often tack a *mono-* prefix to the names of the first four ions in Table 6.5, text p. 152. The prefix is not necessary, and should not be used. The *mono-* prefix may be used for the HPO_4^{2-} ion, the last ion in the table, to differentiate between that ion and the dihydrogen phosphate ion, $H_2PO_4^-$. In that case, the prefix *di-* is essential. With the exception of the two phosphate ions, Greek prefixes are not used in modern inorganic nomenclature.

We've been telling you to learn the system. Where do you find this system? It's all summarized in Table 6.8, text p. 160. Everything you need is here. With this knowledge and a periodic table you can write the names and formulas of thousands of compounds. After learning the system, practice, practice, practice. Tables 6.12 and 6.13 on text pp. 167, 168 offer superb opportunities for practice

The answers in the text on pp. A.18–20 all have the first letter in each answer capitalized. This is for grammatical, not chemical reasons. When you write a chemical name, there is no need to capitalize the first letter unless grammar dictates it should be a capital letter.

Chapter 6 Sample Test

Instructions: For each name given below, write the formula; for each formula given, write the name. Indicate oxidation states only when necessary. You may use a "clean" periodic table.

bromine _____ H_2 _____

nitrogen _____ O_2 _____

chlorine _____ F_2 _____

magnesium oxide _____ CaI_2 _____

sodium fluoride _____ Ba^{2+} _____

aluminum nitride _____ Na_2S _____

calcium phosphide _____ Cl^- _____

potassium bromide _____ Li_2O _____

barium sulfide _____ $AlCl_3$ _____

lithium fluoride _____ Mg_3P_2 _____

manganese(III) ion _____ Cu^{2+} _____

phosphorus tribromide _____ Na_2S _____

perchloric acid _____ S_2F_2 _____

ammonium phosphate _____ H_2S _____

iron(II) nitrate _____ HIO_3 _____

potassium bromite _____ Na_2TeO_3* _____

sodium hydrogen carbonate _____ $Mg(H_2PO_4)_2$ _____

sulfuric acid _____ $PbCl_2$ _____

hydrofluoric acid _____ K_2CO_3 _____

potassium iodide _____ Si_2F_6 _____

oxygen difluoride _____ CuI _____

magnesium sulfate 7-hydrate _____ $BaCl_2 \cdot 2\ H_2O$ _____

* Te is the symbol for tellurium, Z = 52.

Check your sample test answers with those on SG p. 207.

CHAPTER 7

Chemical Formula Problems

Assignment 7A: Chemical Formulas, Formula Masses

A chemical formula tells us the number of each atom that make up a given compound. These atoms have a mass given in atomic mass units, abbreviated amu.

In Section 2.2 (text p. 18) you learned that pure substances are either elements or compounds. In Section 5.5 (text p. 123) you learned that relative masses of elements are given in amu. In this assignment you will learn to find the relative masses of compounds, again using the atomic mass unit.

Look for these big ideas:

1) A **chemical formula** tells how many atoms of each element are present in the formula unit of a substance. The number of each atom is given by a subscript following the symbol of that atom.

2) Like atomic mass of elements, **molecular mass**, or **formula mass**, of compounds is measured in atomic mass units using carbon-12 as the reference standard.

3) To find the formula mass of a compound, add up the atomic masses of its atoms.

Learning Procedures
Study
Sections 7.1–7.2, text pp. 171–175. Focus on Performance Goals 7A–7C as you study.

Answer
Questions and Problems 1–8, text p. 194. Check your answers with those on text p. A.20.

Take
the skills quiz below and on the next page. Check your answers with those on SG pp. 57–58.

Assignment 7A Skills Quiz

1) What is the number of atoms of each element in one molecule of acetic acid, $HC_2H_3O_2$?

2) Write the formula for a compound if its formula unit contains one sulfur atom and three oxygen atoms.

3) What is atomic mass, and in what units is it measured?

4) What is formula mass, and in what units is it measured?

5) What is the formula mass of nickel carbonate, $NiCO_3$?

6) Adrenaline is the molecule that triggers the "fight or flight" reflex in us. The formula of adrenaline is $C_9H_{13}NO_3$. What is the molecular mass of an adrenaline molecule?

Assignment 7B: The Mole, Avogadro's Number and Molar Masses

The atomic mass unit, 1.66×10^{-24} gram, is far too small to detect with even the most sensitive balance. To chemists, the term that suggests a weighable amount of something is the mole, abbreviated mol. A mole of any substance contains 6.02×10^{23} units of that substance. The experimentally derived number 6.02×10^{23} is called Avogadro's number. *Understanding the mole, and being able to use it effectively, is probably the most important single skill you can acquire in this course.*

Look for the big ideas in this assignment:

1) One **mole** of anything contains the same number of objects as the number of atoms in exactly 12 grams of carbon-12. This experimentally determined value is **Avogadro's number, 6.02×10^{23}**.

2) The **mass in grams of one mole** of a substance is called the **molar mass** of that substance.

3) The **molar mass** of a **compound** is the **sum of the molar masses of the atoms** that make up a formula unit of that compound.

4) There are conversion factors between grams of a substance, number of particles of that substance and number of moles of that substance.

There are two important equations used in this assignment. Equation 7.2, text p. 175, may be used as a dimensional analysis conversion factor. This factor enables you to convert between moles and number of units, exactly as you convert between dozens of eggs and number of eggs. Equation 7.3, text p. 177, gives a crucial conversion relationship. It enables you to convert between moles and grams of a substance. This conversion is extremely important, for it is the link between the unseen world of the molecule and the easily seen world of the laboratory or production plant. Learn these conversions well, for you will use them often.

Learning Procedures
Study
Sections 7.3–7.5, text pp. 175–181. Focus on Performance Goals 7D–7H as you study.

Answer
Questions and Problems 9–42, text pp. 194–196. Check answers with those on text pp. A.20–22.

Take
the skills quiz below and on the next page. Check your answers with those on SG p. 58.

Assignment 7B Skills Quiz

1) Distinguish between one mole of a substance and the molar mass of that substance.

2) Criticize, and correct, if wrong, the following statement: "Atomic mass is commonly expressed with units of grams per atom."

3) Calculate the molar masses of carbon dioxide, CO_2, and aluminum phosphate, $AlPO_4$.

4) What is the mass of 0.800 mol sodium hydroxide, NaOH?

5) How many molecules are in 6.00 g of methane, CH_4?

6) What is the mass of 12,000,000,000,000,000,000,000 atoms of lead? Assume three significant figures.

7) What is the number of moles of sodium carbonate, Na_2CO_3, in 225 grams?

Assignment 7C: Percentage Composition, Simplest Formulas and Molecular Formulas

In Assignment 7B you learned to calculate the molar mass of a compound. Because compounds conform to the Law of Definite Composition, the molar mass and the numbers used to calculate it may be used to find the percentage composition of the elements in that compound.

After learning percentage composition, you will use percentage composition calculations to determine the simplest, or empirical, formula of a compound.

Simplest formula? Empirical formula? Let's try a formula problem of a different sort, to explain.

I have some cats. Of these cats, 50% are neutered female, 50% are neutered male. How many cats do I have?

You can't answer that question without more information. All you know is that the ratio of female to male cats is 1:1. This ratio is the simplest formula of my cats. I might have a total of 2 cats, 1 female, 1 male; I might have 6 cats, 3 female, 3 male; I might have 400 cats, 200 female, 200 male.

A calculation from percentage composition (of cats or elements) can only give the simplest formula, or ratio between the elements (or cats) present. You need more information to determine the actual molecular formula. You will be given that information when it is needed in this assignment.

The main ideas in Assignment 7C are:

1) The molar mass and the chemical formula of a compound are used to determine the percentage by mass of each element in that compound.

2) The **simplest formula** gives the simplest ratio, or the ratio in lowest terms, of the elements in a compound. In textbooks in later courses you may find that another name for simplest formula is **empirical formula**.

3) Simplest formulas are calculated from percentage composition data. They are also found from the mass of each element in a sample of a compound.

4) Simplest formulas may or may not be the actual molecular formulas of compounds. Additional information is needed to determine molecular formulas from simplest formulas.

Learning Procedures
Study
Sections 7.6–7.8, text pp. 181–190. Focus on Performance Goals 7I–7L as you study.

Answer
Questions and Problems 43–68, text p. 196. Check your answers with those on text pp. A.22–23.

Take
the skills quiz below and on the next page. Check your answers with those on SG p. 59.

Assignment 7C Skills Quiz

Instructions: Solve each problem in the space provided.

1) What is the percentage composition of aluminum nitrate, $Al(NO_3)_3$?

2) What mass of potassium is present in 53.1 g of potassium phosphate, K_3PO_4?

3) Which of the following could only be empirical formulas? C_2H_6; C_3H_6; C_3H_8O; CH_5N. Explain your answers.

4) An oxide of sulfur is 40.1% sulfur and 59.9% oxygen by mass. Find the simplest formula of this compound.

5) A 5.38 g sample of a pure substance contains 2.35 g of phosphorus, with the remainder oxygen. The molar mass of this substance is 142 g/mole; find the molecular formula of this substance.

Answers to Chapter 7 Skills Quiz Questions

Assignment 7A

1) One molecule of acetic acid, $HC_2H_3O_2$, contains two carbon atoms, four hydrogen atoms and two oxygen atoms. See Example 7.1, text p. 172.

2) The compound with one sulfur atom and three oxygen atoms is SO_3, sulfur trioxide. See Questions and Problems 1 and 2 on text p. 194.

3) Atomic mass is the average mass of the atoms of an element compared to the mass of an atom of carbon-12, which is defined as exactly 12 atomic mass units. The atomic mass unit is the unit of atomic mass. This is a review question; go back to text p. 121.

4) Formula mass is the sum of the atomic masses (in amu) of all the atoms in one formula unit of a substance. Check page 172 in the text.

5) $NiCO_3$: 1(58.7 amu) + 1(12.0 amu) + 3 (16.0 amu) = 118.7 amu   Check Example 7.3, text p. 174.

6) $C_9H_{13}NO_3$: 9(12.01 amu) + 13(1.008 amu) + 1(14.0 amu) + 3(16.0 amu) = 183.2 amu
Study Example 7.4, text p. 175. Check the rules for significant figures and formula mass on text p. 173.

Assignment 7B

1) One mole of a substance is that quantity that has exactly the same number of atoms, molecules or formula units as the number of atoms in exactly 12 grams of carbon-12. Molar mass is the mass in grams of one mole of any substance. Read text pp. 175–177, and Quick Check 7.2 on text p. 178.

2) Atomic mass is not given in grams per atom. The atomic mass of a hydrogen atom, expressed as grams per atom, is 1.66×10^{-24}, an inconvenient number to say the least. Atomic mass is expressed in atomic mass units, which are easier to use. To make certain you know the correct units for atomic mass review text p. 121 and Question and Problem 27 on text p. 132.

3) CO_2: 12.0 + 2(16.0) = 44.0 g/mol
$AlPO_4$: 27.0 + 31.0 + 4(16.0) = 122.0 g/mol   Read text pp. 178–181, Example 7.6.

4) GIVEN: 0.800 mol NaOH   WANTED: g NaOH   PATH: mol NaOH → g NaOH
FACTOR: 40.0 g/mol NaOH $0.800 \text{ mol NaOH} \times \dfrac{40.0 \text{ g NaOH}}{1 \text{ mol NaOH}} = 32.0 \text{ g NaOH}$
See Example 7.7, text p. 179.

5) CH_4: 12.0 g/mol C + 4(1.0 g/mol H) = 16.0 g/mol CH_4
GIVEN: 6.00 g CH_4 WANTED: molecules of CH_4
PATH: g CH_4 → mol CH_4 → molecules CH_4 FACTORS: 16.0 g/mol CH_4, 6.02×10^{23} molecules/mol
$6.00 \text{ g CH}_4 \times \dfrac{1 \text{ mol CH}_4}{16.0 \text{ g CH}_4} \times \dfrac{6.02 \times 10^{23} \text{ molecules CH}_4}{1 \text{ mol CH}_4} = 2.26 \times 10^{23}$ molecules CH_4
Check Example 7.9, text p. 180.

6) GIVEN: 12,000,000,000,000,000,000,000 = 1.20×10^{22} Pb atoms   WANTED: g Pb
PATH: Pb atoms → mol Pb → g Pb   FACTORS: 207.2 g/mol Pb; 6.02×10^{23} atoms/mol
$1.20 \times 10^{22} \text{ Pb atoms} \times \dfrac{1 \text{ mol Pb}}{6.02 \times 10^{23} \text{ Pb atoms}} \times \dfrac{207.2 \text{ g Pb}}{1 \text{ mol Pb}} = 4.13 \text{ g Pb}$
See Example 7.10, text p. 181. Trouble with the exponential notation? Go back to pp. 36–39 in the text.

7) Na_2CO_3: 2(23.0 g/mol Na) + 12.0 g/mol C + 3(16.0 g/mol O) = 106.0 g/mol Na_2CO_3
GIVEN: 225 g Na_2CO_3 WANTED: mol Na_2CO_3 PATH: g Na_2CO_3 → mol Na_2CO_3
FACTOR: 106.0 g/mol Na_2CO_3
$225 \text{ g Na}_2\text{CO}_3 \times \dfrac{1 \text{ mol Na}_2\text{CO}_3}{106.0 \text{ g Na}_2\text{CO}_3} = 2.12 \text{ mol Na}_2\text{CO}_3$
See Example 7.8 text p. 179.

Assignment 7C

1) $Al(NO_3)_3$: 213.0 g/mol

Element	Grams	Percent
Al	$1 \times 27.0 = 27.0$ g Al	$\frac{27.0}{213.0} \times 100 = 12.7\%$ Al
N	$3 \times 14.0 = 42.0$ g N	$\frac{42.0}{213.0} \times 100 = 19.7\%$ N
O	$9 \times 16.0 = 144.0$ g O	$\frac{144.0}{213.0} \times 100 = 67.6\%$ O

Read text pp. 181–183 and see Examples 7.11 and 7.12.

2) GIVEN: 53.1 g K_3PO_4 WANTED: g K   PATH: g $K_3PO_4 \rightarrow$ g K
FACTORS: 212.3 g K_3PO_4/ mol K_3PO_4; 3×39.1 g K/ 212.3 g K_3PO_4; 117.3 g K/ 212.3 g K_3PO_4

$$53.1 \text{ g } K_3PO_4 \times \frac{117.3 \text{ g K}}{212.3 \text{ g } K_3PO_4} = 29.3 \text{ g K}$$ See Examples 7.13–7.15, text pp. 183–184.

3) Only C_3H_8O and CH_5N could be simplest formulas. Simplest formulas are like fractions expressed in lowest terms; you can't reduce the numbers in the ratio further. See text p. 185 for more details.

4)

Element	Grams	Moles	Mole Ratio	Formula Ratio	Simplest Formula	Molecular Formula
S	40.1	$\frac{40.1}{32.1} = 1.25$	1.00	1		
O	59.9	$\frac{59.9}{16.0} = 3.74$	2.99	3	SO_3	

5)

Element	Grams	Moles	Mole Ratio	Formula Ratio	Simplest Formula	Molecular Formula
P	2.35	$\frac{2.35}{31.0} = 0.0758$	1.00	2		
O	3.03	$\frac{3.03}{16.0} = 0.189$	2.50	5	P_2O_5	P_2O_5

For Problem 4, read text pp. 185–187, and review Examples 7.17–7.18, text pp. 187–189. For Problem 5, see text pp. 170-171, and Example 7.19 text p. 190.

Sage Advice and Chapter Clues

A review of Chapter 7 is actually a review of the entire course so far. Chapter 7 ties together what you learned in Chapter 3 about exponential notation, measurements and calculations, in Chapter 5 about the atom and atomic mass, and in Chapter 6 about chemical formulas. The key to understanding

Chapter 7 is understanding the idea of the mole. The mole is the "chemist's dozen," a handy bulk measure of a substance. Because you can use the mole as a conversion relationship between the huge number of particles in a sample and the mass of the sample, the mole is our handle on laboratory-sized chemical mass relationships.

The calculator procedures in Table 7.1, text p. 173 are useful for adding up atomic masses to get a molecular mass. Here's another way to do the same addition, if your calculator has a Σ+ or an M+ key. As you calculate the molar mass of each type of atom in the compound, store that number in a memory register by pressing Σ+ or M+. When you need the molecular mass, press the key that displays the contents of that memory register. Make sure that memory register is cleared before you do this calculation. Find those calculator instruction books yet?

The simplest formula of a compound comes from mass ratio measurements. Like a fraction in lowest terms, these ratios cannot be reduced further. The simplest formula may not be the molecular formula; make sure you can tell them apart. These simplest formulas are also called empirical formulas because the word empirical means "derived from experiment or observation," and these formulas come from mass data obtained by laboratory analysis.

Don't forget the summary Section 7.7 on text p. 184; there's a lot of useful facts there. If you understand that summary, you won't get atomic mass, molecular mass and molar mass confused.

Chapter 7 Sample Test

Instructions: Pick the letter of the *best* choice answer for questions 3 and 4; solve the other problems in the space provided. You may use a "clean" periodic table.

1) How many nitrogen atoms and how many oxygen atoms are in one molecule of dinitrogen pentoxide, N_2O_5?

2) There are one phosphorus atom and five fluorine atoms in one molecule of phosphorus pentafluoride. Write the formula for this compound.

3) Identify the *incorrect* statement among the following:
   (a) A mole is that quantity of a substance that contains 6.02×10^{23} particles of that substance.
   (b) One mole of any substance contains the same number of particles as one mole of any other substance.
   (c) One mole of any substance contains the same number of particles as the number of atoms in exactly 12 grams of carbon-12.
   (d) A mole is a quantity of a substance that has a mass of exactly 12 grams.

4) Identify the *correct* statement among the following:
   (a) The molar mass of atoms of an element is numerically equal to its atomic mass.
   (b) The molar mass is the mass in amu of one mole of any substance.
   (c) Molar mass and atomic mass are always numerically equal.
   (d) Molar masses of compounds are always equal to atomic masses of those compounds.

5) Calculate the molar mass of potassium chloride, KCl.

6) What is the mass of 1.06×10^{24} molecules of carbon monoxide?

7) What mass of sodium nitrate, $NaNO_3$, contains 1.62 g of nitrogen?

8) How many moles of potassium sulfate, K_2SO_4, are in 37.5 grams?

9) Calculate the percentage composition of barium hydroxide, $Ba(OH)_2$.

10) From the following, pick those that are simplest formulas: $C_{10}H_6$; $C_2H_6O_2$; CH_3; C_2H_2; $C_{30}H_{50}O$.

11) A compound has the percentage composition 92.25% C and 7.75% H. Calculate the simplest formula for this compound. If the molar mass of this compound is 78 g/mol, calculate the true molecular formula for this compound.

Check your sample test answers with those on SG pp. 207–208.

CHAPTER 8
Chemical Reactions and Equations

Assignment 8A: Balancing Chemical Equations

In Section 2.2 (text p. 23) you learned about the characteristics of a chemical change. The first characteristic (PG 2I) was the chemists describe chemical changes, or reactions, by writing chemical equations. In this assignment you'll learn to describe the chemical reactions of elements and compounds by using a chemical equation.

The main ideas in this assignment are:

1) A **chemical equation** is a **shorthand description** of a chemical reaction.

2) Chemical equations can be interpreted on a molecular scale or a molar scale.

4) A balanced chemical equation reflects the Law of Conservation of Mass.

5) The subscripts in a chemical equation may *never* be changed simply to balance the equation.

Changing a subscript in the formula of a substance changes the chemical identity of that substance (Law of Definite Composition.)

6) When balancing an equation start big and work to small.

Learning Procedures
Study
Sections 8.1–8.2, text pp. 199–201. Focus on Performance Goal 8A as you study.

Answer
Equations 1–24 from the Equation Balancing Exercise, text p. 219. Check your answers with those on text p. A.23.

Take
the skills quiz below and on the next page. Check your answers with those on SG p. 66.

Assignment 8A Skills Quiz

Instructions: Balance the following chemical equations, for which correct formulas are already written.

$$CH_4 + O_2 \rightarrow CO_2 + H_2O$$

$LiCl + AgNO_3 \rightarrow LiNO_3 + AgCl$

$H_2 + O_2 \rightarrow H_2O$

$H_2SO_4 + NaOH \rightarrow Na_2SO_4 + H_2O$

$PCl_3 + H_2O \rightarrow H_3PO_3 + HCl$

$Pb(NO_3)_2 + HBr \rightarrow PbBr_2 + HNO_3$

$Al + HCl \rightarrow AlCl_3 + H_2$

Assignment 8B: Writing and Balancing Chemical Equations

In the last assignment you got a start on balancing chemical equations, if all the reactants and products were given to you. In this assignment you'll learn to write the equations yourself, before balancing them.

Scientists try to understand the universe by organizing numerous experimental results into simple, logical patterns. Dalton did this with his atomic model, as did Mendeleev and Meyer with the periodic table. In this assignment you will learn to describe the chemical reactions of elements and compounds by using a chemical equation, and to organize these equations by types of reactants and products.

The important ideas in this assignment are:

1) You must know the correct formulas for all the reactants and products before you can write a chemical equation.

2) Chemical equations are organized by **reaction types**; these types include **combination** (putting things together), **decomposition** (taking things apart) and **complete oxidation** (burning) reactions.

Learning Procedures
Study
Sections 8.3–8.6, text pp. 205–210, top. Focus on Performance Goals 8B–8D as you study.

Answer
Questions and Problems 1–14, text pp. 219–220. Check your answers with those on text p. A.24.

Take
the skills quiz below and on the next page. Check your answers with those on SG p. 66.

Assignment 8B Skills Quiz

1) Write the equation for the formation of hydrogen chloride, HCl, gas by direct combination of its elements, both of which are also gases.

2) Elemental sulfur reacts with oxygen gas to form sulfur trioxide, SO_3. Write the balanced equation for this reaction.

3) When heated, solid magnesium carbonate decomposes into solid magnesium oxide and carbon dioxide gas. Write the equation for this reaction.

4) Write the equation for the complete burning of liquid methanol, CH_3OH.

Assignment 8C: Reactions in Water Solution

In Assignment 8A, you learned to balance chemical equations. In this assignment, you will learn to recognize oxidation-reduction, precipitation and neutralization reactions. Oxidation-reduction reactions may be described by single replacement equations while precipitation and neutralization reactions may both be described by double replacement equations. These equation types are important because they describe the major reactions that occur in water solutions. Water is the reaction medium you will most often encounter. Look for these big ideas:

1) Many reactions in water solution are described by either a single replacement equation or by a double replacement equation.

2) An important oxidation-reduction reaction has the single replacement equation of the general form **acid + metal $\rightarrow$ H$_2$ + salt**.

3) An important reaction of acids is a **neutralization reaction**. This reaction is depicted by a double replacement equation having the general form **acid + base $\rightarrow$ water + salt**.

Learning Procedures

Study
Sections 8.7–8.10, text pp. 212–216. Focus on Performance Goals 8E–8G as you study.

Answer
Questions and Problems 15–28, text p. 220. Don't forget about Questions and Problems 29–70, either. Check your answers with those on text p. A.24.

Take
the skills quiz below and on the next page. Check your answers with those on SG p. 66.

Assignment 8C Skills Quiz

1) Zinc metal is placed in aqueous sulfuric acid and a reaction occurs. Identify the reaction type and write the equation for this reaction.

2) Aqueous solutions of sodium iodide and lead nitrate are mixed and solid lead iodide forms. Identify the reaction type and write the equation for this reaction.

3) Identify the reaction type and write the equation for the reaction between potassium hydroxide and nitric acid in water solution.

Answers to Chapter 8 Skills Quiz Questions

Assignment 8A

1) $CH_4 + 2\ O_2 \rightarrow CO_2 + 2\ H_2O$ $LiCl + AgNO_3 \rightarrow LiNO_3 + AgCl$
   $2\ H_2 + O_2 \rightarrow 2\ H_2O$ $H_2SO_4 + 2\ NaOH \rightarrow Na_2SO_4 + 2\ H_2O$
   $PCl_3 + 3\ H_2O \rightarrow H_3PO_3 + 3\ HCl$ $Pb(NO_3)_2 + 2\ HBr \rightarrow PbBr_2 + 2\ HNO_3$
   $2\ Al + 6\ HCl \rightarrow 2\ AlCl_3 + 3\ H_2$

   Check text pp. 201–204, and Examples 8.1–8.3. If you haven't done all of the Equation-Balancing Exercise on text p. 219, do it now! Practice, practice....

Assignment 8B

1) $H_2(g) + Cl_2(g) \rightarrow 2\ HCl(g)$ See Examples 8.4–8.6, text pp. 205–206.
2) $2\ S(s) + 3\ O_2(g) \rightarrow 2\ SO_3(g)$ See Example 8.5, text p. 206.
3) $MgCO_3(s) \xrightarrow{\Delta} MgO(s) + CO_2(g)$ See Example 8.5, text p. 208.
4) $2\ CH_3OH(\ell) + 3\ O_2(g) \xrightarrow{\Delta} 2\ CO_2(g) + 4\ H_2O(g)$ See text pp. 208–209, Example 8.11.

   If you forgot which elements are diatomic, go back to Figure 6.1 on text p. 138.

Assignment 8C

1) One reactant is a metal, the other an acid. This is a redox reaction with a single replacement equation, the metal replacing hydrogen:
   $Zn(s) + H_2SO_4(aq) \rightarrow ZnSO_4(aq) + H_2(g)$ See text pp. 210–212, and Example 8.13.

2) Both reactants are compounds; neither reactant is an acid nor a base. This is a precipitation reaction described by a double replacement equation:
   $2\ NaI(aq) + Pb(NO_3)_2(aq) \rightarrow PbI_2(s) + 2\ NaNO_3(aq)$ See text pp. 214–216, Examples 8.15, 8.16.

3) **KOH** is a base; **HNO_3** is an acid; this is a neutralization reaction, a case of a double replacement equation: $KOH(aq) + HNO_3(aq) \rightarrow KNO_3(aq) + HOH(\ell)$
   Check text pp. 214–216, particularly Examples 8.17 and 8.18.

Sage Advice and Chapter Clues

When writing chemical equations, you may encounter several minor pitfalls. Fortunately, they are easily avoided. The most common, and dreaded, error is not following the direct three-step approach given on page 205 in the textbook.

First, look at the reactants and conditions to classify the reaction. The major reaction types that do not occur in water are combination, decomposition and complete oxidation. (Don't forget to add O_2 in burning reactions.) In water solution, some reaction types are oxidation-reduction, and precipitation and neutralization reactions. Oxidation-reduction reactions are described by single replacement equations; precipitation (not percipitation) and neutralization reactions are described by double replacement equations. Remember, if you can recognize the reaction type, you can predict the products and write correct formulas for them. Table 8.1, text p. 217, will help you develop this skill.

After you have the correct chemical formula for each reactant and product, balance the number of atoms of each element on each side of the equation by changing coefficients only. Never, **Never NEVER**, change the subscript to balance an equation. A coefficient is not the same as a subscript; 3 Br_2 is not the same as Br_6 or 6 Br. The formula writing skills you developed in Chapter 6 will serve you well here.

It's easier to balance double displacement equations if you balance groups rather than atoms. If you see a polyatomic group that's the same on both sides of the equation, balance the number of that group. For example, if there are sulfate groups on both sides of the arrow, there must be the same number of sulfates Don't balance all the S atoms in the sulfates first, then all the O atoms in the sulfates. That takes much too long. Just balance the sulfates.

In Chapter 6 you learned the formula for water is H_2O. That's true, but water often reacts as if it were HOH. You may find acid-base neutralization equations easier to balance if you write the water product as HOH. All the H atoms at the front of the water come from the reactant acid, all the OH comes from the reactant base, and the number of H equals the number of OH.

You must be able to balance these chemical equations, because in Chapter 9 you must start with a balanced chemical equation to see how much of a reactant or product is involved in a reaction.

Chapter 8 Sample Test

Instructions: In the spaces provided, write a chemical equation for each reaction described. You may use a "clean" periodic table on this test.

1) Solid calcium oxide is formed from its elements.

2) Barium oxide and water result from the decomposition of barium hydroxide.

3) Gaseous C_4H_9CHO is completely oxidized.

4) Hydrogen gas and a solution of lithium hydroxide are the products of the reaction between lithium metal and water.

5) Potassium hydroxide solution reacts with a solution of copper(II) nitrate, $Cu(NO_3)_2$. Copper(II) hydroxide, $Cu(OH)_2$, is a product.

6) A water solution of sulfuric acid reacts with a water solution of sodium hydroxide.

7) When hydrochloric acid is poured over solid calcium carbonate, carbon dioxide bubbles off, leaving water and aqueous calcium chloride as the other products.

Check your sample test answers with those on SG p. 208.

CHAPTER 9

Quantity Relationships in Chemical Reactions

Assignment 9A: Mass Stoichiometry

The term **stoichiometry** refers to the quantitative relationships between the substances involved in a chemical reaction. Chemical quantities may be measured in several ways. In this assignment, you will consider only grams. Later in this chapter, you will learn to use thermochemical energy as a stoichiometric quantity. The method of stoichiometry is the same regardless of the quantities used; a single principle covers the entire area of reaction quantities. In this assignment, you are directed to the *method* of solving a stoichiometry problem, rather than to a specific problem.

New ideas that are introduced in this assignment are:

1) The **coefficients** in a chemical equation express the **mole relationships** between the different substances in the reaction. The coefficients may be used in a dimensional analysis conversion from moles of one substance to moles of another.

2) The "stoichiometry pattern" on text p. 225 is a three-step method that can be used to solve almost all stoichiometry problems.

Item 2 is extremely important. The stoichiometry pattern is not difficult to learn; once learned, it is easy to apply. Learn the pattern well in this assignment using mass as the only quantity measurement. If you have learned the pattern in this assignment, Assignment 9C becomes simple. You will already know how to solve stoichiometry problems involving thermochemical energy in Assignment 9C, because the energy problems also fit into the stoichiometry pattern of Assignment 9A.

Learning Procedures
Study
Sections 9.1–9.2, text pp. 223–229. Focus on Performance Goals 9A–9B as you study.

Answer
Questions and Problems 1–30, text pp. 247–249. Check your answers with those on text pp. A.24, 25.

Take
the skills quiz on the next page. Check your answers with those on SG p. 74.

Assignment 9A Skills Quiz

1) How many moles of nitric acid will result if 1.80 moles of NO are produced in the reaction below?
$$3\ NO_2(g) + H_2O(g) \rightarrow 2\ HNO_3(g) + NO(g)$$

2) Find the number of grams of water required for the reaction of 0.912 moles of NO_2 in the equation in Question 1, above.

3) How many grams of calcium fluoride, CaF_2, will precipitate from a combination of two solutions, one containing 27.9 grams of sodium fluoride, NaF and the other an excess of calcium nitrate, $Ca(NO_3)_2$? (You must first write the chemical equation for the reaction.)

4) Sulfuric acid, H_2SO_4, is the highest volume product of the chemical industry. Production for 1993 is estimated as 411 gigamoles (Gmol). Refer to Table 3.2, text p. 49 for metric prefixes. How many grams of oxygen are needed to produce this amount of H_2SO_4 in the reaction below?
$$2\ S(s) + 2\ H_2O(g) + O_2(g) \rightarrow 2\ H_2SO_4(g)$$

Assignment 9B: Percent Yield and Limiting Reactant Problems

In Assignment 9A, you learned the fundamental theory that underlies all quantitative reaction chemistry. But it **is** theory, and practice more often than not does not exactly conform to theory.

The mass of product actually obtained in a reaction is not always the same as the mass calculated by stoichiometry. Only rarely, and only under carefully controlled conditions are two or more reactants brought together in just the right quantities so that all reactants are totally consumed. Usually, there is more than enough of one substance (usually the cheapest) to react with all of another. The reaction proceeds until all of the second substance is used up. In this assignment, you add the ideas of percent yield and limiting reactant to stoichiometric calculations.

Look for these critical ideas.

1) The **mass efficiency** of a reaction is stated in **percent yield**, in which the actual yield is expressed as a percent of the theoretical yield calculated by stoichiometry.

2) The **limiting reactant** determines the maximum amount of product that can be obtained in a chemical reaction.

Learning Procedures

Study
Sections 9.3–9.4, text pp. 229–241. Focus on Performance Goals 9C–9D as you study.

Answer
Questions and Problems 31–50, text pp. 249–250. Check answers with those on text pp. A.25–26.

Take
the skills quiz below and on the next page. Check your answers with those on SG pp. 75–77.

Assignment 9B Skills Quiz

1) Calculate the theoretical yield of NO and the percentage yield of NO if 49.2 grams of NO_2 yield 8.90 grams of NO in the following reaction:
$$3\ NO_2(g) + H_2O(g) \rightarrow 2\ HNO_3(g) + NO(g)$$

2) How many grams of NO_2 are required to produce 1.25 kilograms of nitric acid in the reaction in Question 1 if the percent yield is 73.8%?

3) Calcium metal, 6.43 grams, is placed into a solution that contains 8.98 grams of HCl. Assume the reaction proceeds until the limiting reactant is consumed. (You must write the reaction equation to solve this problem.)
   (a) What mass of calcium chloride, $CaCl_2$, and what mass of hydrogen gas, H_2, will be produced?
   (b) Identify the reactant that is in excess and calculate the number of grams of that reactant that will remain unreacted.

4) In the reaction $2\ AgNO_3(aq) + H_2S(aq) \rightarrow Ag_2S(s) + 2\ HNO_3(aq)$, 28.1 grams of silver nitrate and 3.50 grams of H_2S react until the limiting reactant is exhausted. Calculate the number of grams of silver sulfide that will precipitate. Also, identify the reactant initially in excess, and calculate the number of grams that remain unreacted.

Assignment 9C: Energy and Thermochemical Stoichiometry

In Assignment 9A, we promised that if you learned how to use the stoichiometry pattern in solving problems involving mass, you would also know how to solve problems with other quantity units. Here's where you cash in on your previous knowledge. In this assignment, you'll learn to apply the stoichiometry pattern to problems involving heat energy.

The major ideas in this assignment are:

1) The SI **unit of energy** is the **joule**, a force of one newton applied for a distance of one meter.

2) Another energy unit used by chemists is the **calorie**, which is equal to **4.184 joules**.

3) The joule (J) and the calorie (cal) are small energy units, so units 1000 times larger, the **kilojoule, kJ**, and the **kilocalorie, kcal**, are often used.

4) The amount of heat given off or absorbed in a chemical reaction is called the **change of enthalpy**, symbolized by ΔH.

5) The ΔH of a reaction may be included in the chemical equation as a reactant or a product, or it may be written next to the equation. The equation is then called a **thermochemical equation**.

6) For an **endothermic** reaction, **ΔH is positive; heat is a reactant** in the thermochemical equation. For an **exothermic** reaction, **ΔH is negative; heat is a product** in the thermochemical equation.

7) Using the ΔH of a thermochemical equation and the coefficient of any substance in that equation, you can convert in either direction from moles of that substance to amount of heat absorbed or produced.

8) In thermochemical equations the value of ΔH may be given in either kJ or in kJ/mole.

Learning Procedures

Study
Sections 9.5–9.7, text pp. 242–245. Focus on Performance Goals 9E–9G as you study.

Answer
Questions and Problems 51–70, text pp. 250–251. Check your answers with those on text p. A.26.

Take
the skills quiz below and on the next page. Check your answers with those on SG p. 77.

Assignment 9C Skills Quiz

Instructions: Solve each problem in the space provided. A correct setup, beginning with the given quantity, is required for a correct answer when dimensional analysis is used.

1) Express 149 calories as joules and as kilojoules.

2) Express 342 joules as calories and as kilocalories.

3) The formation of two moles of ammonia gas from nitrogen gas and hydrogen gas is a reaction for which ΔH = -184 kJ. Write this thermochemical equation in two different ways.

4) Calculate the heat flow if 148 g of $H_2O(g)$ form in the reaction
$$2 H_2(g) + O_2(g) \rightarrow 2 H_2O(g) \qquad \Delta H = -572 \text{ kJ}$$

Answers to Chapter 9 Skills Quiz Questions

Assignment 9A

1) GIVEN: 1.80 mol NO   WANTED: mol HNO_3 PATH: mol NO → mol HNO_3
   FACTOR: 2 mol HNO_3/ 1 mol NO
   $$1.80 \text{ mol NO} \times \frac{2 \text{ mol HNO}_3}{1 \text{ mol NO}} = 3.60 \text{ mol HNO}_3$$
   Read text pp. 223–224, Examples 9.1 and 9.2.

2) GIVEN: 0.912 mol NO_2 WANTED: g H_2O PATH: mol NO_2 → mol H_2O → g H_2O
   FACTORS: 1 mol H_2O/ 3 mol NO_2, 18.0 g H_2O/ mol H_2O
   $$0.912 \text{ mol NO}_2 \times \frac{1 \text{ mol H}_2\text{O}}{3 \text{ mol NO}_2} \times \frac{18.0 \text{ g H}_2\text{O}}{1 \text{ mol H}_2\text{O}} = 5.47 \text{ g H}_2\text{O}$$
   Text pp. 225–229
   Examples 9.4–9.5

3) EQUATION: 2 NaF + $Ca(NO_3)_2$ → CaF_2 + 2 $NaNO_3$
   GIVEN: 27.9 g NaF   WANTED: g CaF_2
   PATH: g NaF → mol NaF → mol CaF_2 → g CaF_2
   FACTORS: 42.0 g NaF/ mol NaF; 1 mol CaF_2/2 mol NaF; 78.1 g CaF_2/mol CaF_2
   $$27.9 \text{ g NaF} \times \frac{1 \text{ mol NaF}}{42.0 \text{ g NaF}} \times \frac{1 \text{ mol CaF}_2}{2 \text{ mol NaF}} \times \frac{78.1 \text{ g CaF}_2}{1 \text{ mol CaF}_2} = 25.9 \text{ g CaF}_2$$
   Example 9.6
   text p. 228

4) GIVEN: 411 Gmol H_2SO_4 WANTED: g O_2
   PATH: Gmol H_2SO_4 → Gmol O_2 → Gg O_2 → g O_2
   FACTORS: 1 Gmol O_2/ 2 Gmol H_2SO_4; 32 Gg O_2/ Gmol O_2; 10^9 g O_2/ 1 Gg O_2
   $$411 \text{ Gmol H}_2\text{SO}_4 \times \frac{1 \text{ Gmol O}_2}{2 \text{ Gmol H}_2\text{SO}_4} \times \frac{32.0 \text{ Gg O}_2}{1 \text{ Gmol O}_2} \times \frac{10^9 \text{ g O}_2}{1 \text{ Gg O}_2} = 6.58 \times 10^{12} \text{ g O}_2$$
   See Example 9.7, text p. 229.

Assignment 9B

1) GIVEN: 49.2 g NO_2 WANTED: g NO (theo)
   PATH: g NO_2 → mol NO_2 → mol NO → g NO
   FACTORS: 46.0 g NO_2/ mol NO_2; 1 mol NO/ 3 mol NO_2; 30.0 g NO/ mol NO

   $$49.2 \text{ g } NO_2 \times \frac{1 \text{ mol } NO_2}{46.0 \text{ g } NO_2} \times \frac{1 \text{ mol } NO}{3 \text{ mol } NO_2} \times \frac{30.0 \text{ g } NO}{1 \text{ mol } NO} = 10.7 \text{ g NO (theo)}$$

   GIVEN: 8.90 g (act); 10.7 g (theo)   WANTED: % yield
   EQUATION: % yield = $\frac{\text{actual yield}}{\text{theoretical yield}} \times 100 = \frac{8.90 \text{ g}}{10.7 \text{ g}} \times 100 = 83.2\%$ Text pp. 229–230 Example 9.8

2) GIVEN: 1.25 kg = 1.25×10^3 g HNO_3 (act)   WANTED: g HNO_3 (theo)
   PATH: g HNO_3 (act) → g HNO_3 (theo)   FACTORS: 73.8 g HNO_3(act)/ 100 g HNO_3(theo)

   $$1.25 \times 10^3 \text{ g } HNO_3 \text{ (act)} \times \frac{100 \text{ g } HNO_3 \text{ (theo)}}{73.8 \text{ g } HNO_3 \text{ (act)}} = 1.69 \times 10^3 \text{ g } HNO_3 \text{ (theo)}$$

   EQUATION: 3 NO_2(g) + H_2O(g) → 2 HNO_3(g) + NO(g)
   GIVEN: 1.69×10^3 g HNO_3 WANTED: g NO_2
   PATH: g HNO_3 (theo) → mol HNO_3 → mol NO_2 → g NO_2
   FACTORS: 63.0 g HNO_3/ mol HNO_3; 3 mol NO_2/ 2 mol HNO_3; 46.0 g NO_2/ mol NO_2

   $$1.69 \times 10^3 \text{ g } HNO_3 \times \frac{1 \text{ mol } HNO_3}{63.0 \text{ g } HNO_3} \times \frac{3 \text{ mol } O_2}{2 \text{ mol } HNO_3} \times \frac{46.0 \text{ g } NO_2}{1 \text{ mol } NO_2} = 1.85 \times 10^3 \text{ g } NO_2$$

   This problem could also be done as a single dimensional analysis setup,

   $$1.25 \times 10^3 \text{ g } HNO_3 \text{ (act)} \times \frac{100 \text{ g (theo)}}{73.8 \text{ g (act)}} \times \frac{1 \text{ mol } HNO_3}{63.0 \text{ g } HNO_3} \times \frac{3 \text{ mol } NO_2}{2 \text{ mol } HNO_3} \times \frac{46.0 \text{ g } NO_2}{1 \text{ mol } NO_2} = 1.86 \times 10^3 \text{ g}$$

   Whew! Study Example 9.9, text pp. 231–232.

3) The balanced equation for this reaction is given below, above the table.

	Ca	+	2 HCl	→	$CaCl_2$	+	H_2.
Grams at start	6.43		8.98		0		0
Molar Mass (g/mole)	40.1		36.5		111.1		2.02
Moles at start	0.160		0.246		0		0

   If Ca is the limiting reactant, how many moles of HCl are required to react with all the Ca? This calculation is shown below left.

   $0.160 \text{ mol Ca} \times \frac{2 \text{ mol HCl}}{1 \text{ mol Ca}} = 0.320 \text{ mol HCl}$ $0.246 \text{ mol HCl} \times \frac{1 \text{ mol Ca}}{2 \text{ mol HCl}} = 0.123 \text{ mol Ca}$

   The table shows that there are 0.246 mol HCl, not enough to react with all of the calcium. The HCl is therefore the limiting reactant. Let's check this by calculating how many mol of Ca are needed to react with all the HCl, in the next column....

   We need 0.123 mol Ca to react with all the HCl present; because we have 0.160 mol Ca, we have more than enough Ca to react with all the HCl. This is another way of saying that HCl is the limiting reactant. Let's update the table:

	Ca	+	2 HCl	→	CaCl$_2$	+	H$_2$.
Grams at start	6.43		8.98		0		0
Molar Mass (g/mole)	40.1		36.5		111.1		2.02
Moles at start	0.160		0.246		0		0
Moles used(-), produced(+)	-0.123		-0.246		0.123		0.123
Moles at end	0.037		0		0.123		0.123

The question asked us the find the *mass* of CaCl$_2$ and H$_2$ produced, not the moles produced. We'll get these masses by using a calculation like that shown in the *Smaller Amount Method*, text p. 239. Note that we already know that HCl is the limiting reactant.

$$8.98 \text{ g HCl} \times \frac{1 \text{ mol HCl}}{36.5 \text{ g HCl}} \times \frac{1 \text{ mol CaCl}_2}{2 \text{ mol HCl}} \times \frac{111.1 \text{ g CaCl}_2}{1 \text{ mol CaCl}_2} = 13.7 \text{ g CaCl}_2 \text{ produced}$$

Now for the amount of H$_2$ that can be produced, again from the HCl....

$$8.98 \text{ g HCl} \times \frac{1 \text{ mol HCl}}{36.5 \text{ g HCl}} \times \frac{1 \text{ mol H}_2}{2 \text{ mol HCl}} \times \frac{2.02 \text{ g H}_2}{1 \text{ mol H}_2} = 0.248 \text{ g H}_2 \text{ produced}$$

(b) The calcium required to react with all of the HCl is 0.123 mol, from the calculation of the limiting reactant. We started with 0.160 mol of Ca, and reacted 0.123 mol of Ca, leaving 0.037 mol Ca,

$$0.037 \text{ mol Ca} \times \frac{40.1 \text{ g Ca}}{1 \text{ mol Ca}} = 1.5 \text{ g Ca remains unreacted}$$

4) The balanced equation for this reaction is given below, above the table. We'll solve this problem using the *Smaller Amount Method* from text p. 239.
EQUATION: 2 AgNO$_3$(aq) + H$_2$S(aq) → Ag$_2$S(s) + 2 HNO$_3$(aq)
GIVEN: 28.1 g AgNO$_3$, 3.50 g H$_2$S   WANTED: g Ag$_2$S; mass of excess reactant remaining
PATHS: g AgNO$_3$ → mol AgNO$_3$ → mol Ag$_2$S → g Ag$_2$S; g H$_2$S → mol H$_2$S → mol Ag$_2$S → g Ag$_2$S
FACTORS: 169.9 g AgNO$_3$/ mol AgNO$_3$, 2 mol AgNO$_3$/ 1 mol H$_2$S, 247.9 g Ag$_2$S/ mol Ag$_2$S
34.1 g H$_2$S/ mol H$_2$S, 1 mol Ag$_2$S/ 1 mol H$_2$S

Assuming AgNO$_3$(aq) is the limiting reactant, we calculate:

$$28.1 \text{ g AgNO}_3 \times \frac{1 \text{ mol AgNO}_3}{169.9 \text{ g AgNO}_3} \times \frac{1 \text{ mol Ag}_2\text{S}}{2 \text{ mol AgNO}_3} \times \frac{247.9 \text{ g Ag}_2\text{S}}{\text{mol Ag}_2\text{S}} = 20.5 \text{ g Ag}_2\text{S}$$

Assuming H$_2$S(aq) is the limiting reactant, we calculate:

$$3.50 \text{ g H}_2\text{S} \times \frac{1 \text{ mol H}_2\text{S}}{34.1 \text{ g H}_2\text{S}} \times \frac{1 \text{ mol Ag}_2\text{S}}{1 \text{ mol H}_2\text{S}} \times \frac{247.9 \text{ g Ag}_2\text{S}}{\text{mol Ag}_2\text{S}} = 25.4 \text{ g Ag}_2\text{S}$$

So AgNO$_3$ is the limiting reactant, and 20.5 g Ag$_2$S precipitates. So how much H$_2$S remains unreacted? To find the excess, we'll determine how much H$_2$S reacts with the AgNO$_3$, then subtract from the initial mass of H$_2$S.

$$28.1 \text{ g AgNO}_3 \times \frac{1 \text{ mol AgNO}_3}{169.9 \text{ g AgNO}_3} \times \frac{1 \text{ mol H}_2\text{S}}{2 \text{ mol AgNO}_3} \times \frac{34.1 \text{ g H}_2\text{S}}{\text{mol H}_2\text{S}} = 2.82 \text{ g H}_2\text{S } \textit{reacted}$$

So we have left 3.50 - 2.82 = 0.68 g H$_2$S remaining. Let's put all this into a table....

	2 AgNO$_3$(aq) +	H$_2$S(aq) →	Ag$_2$S(s) +	2 HNO$_3$(aq)
Grams at start	28.1	3.50	0	0
Molar Mass (g/mole)	169.9	34.1	247.9	63.0
Moles at start	0.165	0.103	0	0
Moles used(-), produced(+)	-0.165	-0.0825	+0.0825	+0.165
Moles at end	0	0.021	0.0825	0.165
Grams at end	0	0.72	20.5	10.4

Note that if you use moles for determining how much H$_2$S remains, you will have a final subtraction of 0.103 - 0.0825 = 0.021 mol H$_2$S remaining, which gives 0.72 g H$_2$S.

Contemplate Examples 9.16–9.17, text pp. 240–241.

Assignment 9C

1) GIVEN: 149 cal   WANTED: J, kJ   PATHS: cal → J; cal → kJ
   FACTORS: 4.184 J/ cal, 10^3 J/ 1 kJ, 10^3 cal/ 1 kcal

   $$149 \text{ cal} \times \frac{4.184 \text{ J}}{1 \text{ cal}} = 623 \text{ J}$$

   $$149 \text{ cal} \times \frac{4.184 \text{ J}}{1 \text{ cal}} \times \frac{1 \text{ kJ}}{10^3 \text{ J}} = 0.623 \text{ kJ}$$

2) GIVEN: 342 J   WANTED: cal, kcal   PATHS: J → cal; J → kcal
   FACTORS: 4.184 J/ cal, 10^3 J/ 1 kJ, 10^3 cal/ 1 kcal

   $$342 \text{ J} \times \frac{1 \text{ cal}}{4.184 \text{ J}} = 81.7 \text{ cal}$$

   $$342 \text{ J} \times \frac{1 \text{ cal}}{4.184 \text{ J}} \times \frac{1 \text{ kcal}}{10^3 \text{ cal}} = 0.0817 \text{ kcal}$$

   For 1 and 2 read text p. 242 and study Example 9.18.

3) N$_2$(g) + 3 H$_2$(g) → 2 NH$_3$(g)   ΔH = -184 kJ
   N$_2$(g) + 3 H$_2$(g) → 2 NH$_3$(g) + 184 kJ

   Read text p. 242 and study Example 9.19, text pp. 243–244, top.

4) EQUATION: 2 H$_2$(g) + O$_2$(g) → 2 H$_2$O(g) + 572 kJ
   GIVEN: 148 g H$_2$O   WANTED: kJ   PATH: g H$_2$O → mol H$_2$O → kJ
   FACTORS: 18.0 g H$_2$O/ mol H$_2$O; 572 kJ/ 2 mol H$_2$O

   $$148 \text{ g H}_2\text{O} \times \frac{1 \text{ mol H}_2\text{O}}{18.0 \text{ g H}_2\text{O}} \times \frac{572 \text{ kJ}}{2 \text{ mol H}_2\text{O}} = 2.35 \times 10^3 \text{ kJ}$$

   See text pp. 244–245
   Examples 9.20, 9.21.

Sage Advice and Chapter Clues

When the philosopher George Santayana wrote "They who cannot remember the past are condemned to repeat it.", he was referring to the many forgotten lessons of history.

He could also be referring to the lessons of chemistry. Formula writing from Chapter 6, formula calculations from Chapter 7 and equation writing from Chapter 8 all come together in solving stoichiometry problems. If you know these three basic skills well, and can follow the method by which all stoichiometry problems are solved, this chapter will present no difficulty. If you are weak in the basic skills needed to do stoichiometry problems, heal the weakness *before* you go on. You are building a pyramid of skills, with each new skill resting on the foundation of the previous skills. The foundation *must* be firm, or collapse and panic will soon follow. Set the foundation firmly, so you build for the ages.

You were introduced to the units of heat energy, the joule and the calorie. Lots of people wonder at this point how a joule, a force through a displacement, can represent heat. Try this experiment: Place the palms of your hands firmly together, so it would take a force to rub one palm against the other (this rubbing motion is the displacement). Now rub your palms. What do you feel? Right. Heat, a force through a displacement.

A word of caution. The sign of ΔH can cause problems. If ΔH is greater than zero, the reaction is endothermic; heat is a reactant, and a heat term can appear on the left side of the thermochemical equation. If ΔH is less than zero, the reaction is exothermic; heat is a reaction product, and a heat term can appear on the right side of the thermochemical equation.

Both mass and thermochemical types of stoichiometry problems are easily solved using the stoichiometry pattern given on text p. 225. Never mind the details by which problems differ. Apply the pattern consistently and the details become similar, and also much clearer.

Chapter 9 Sample Test

Instructions: Solve each problem in the space provided. You may use a "clean" periodic table.

Questions 1–4 refer to the equation $2\ C_5H_{10} + 15\ O_2 \rightarrow 10\ CO_2 + 10\ H_2O$

1) How many moles of carbon dioxide result from the reaction of 3 moles of C_5H_{10}?

2) What mass of oxygen must react to form 12.0 moles of CO_2?

3) If the reaction consumes 4.12 grams of oxygen, how many grams of water are also produced?

4) The reaction of 6.81 grams of C_5H_{10} yields 19.2 grams of CO_2. Find the percentage yield.

5) Calculate the grams of Fe_2O_3 produced if 15.2 grams of iron and 7.41 grams of oxygen react until one of them is completely consumed in the reaction $4\,Fe + 3\,O_2 \rightarrow 2\,Fe_2O_3$

6) Identify the substance that is in excess in Question 5, and calculate the mass of that substance that remains unreacted.

7) Express 127 joules as kilocalories.

8) Write in two different forms the thermochemical equation showing the decomposition of one mole of $H_2O(g)$ into hydrogen gas and oxygen gas if ΔH for this reaction is +286 kJ.

9) How many grams of C_5H_{10} must be consumed to release 2.22 kJ in the reaction given below?

$$2\ C_5H_{10}(g) + 15\ O_2(g) \rightarrow 10\ CO_2(g) + 10\ H_2O(g) \quad \Delta H = -6.08\ \text{kJ}$$

CHAPTER

10

Atomic Theory and the Periodic Table: A Modern View

We ended Chapter 5 at roughly the midpoint of the 40 year revolution (circa 1890–1930) that completely changed our view of atoms and the universe. In 1890, physics reflected Victorian society. There was a place for everything, and everything was in its place. The universe was composed of matter and energy, and that was that.

Or was it? Rutherford's 1911 model of the atom (text p. 118) placed all the positive charge of the atom, the protons, in a small, dense nucleus, and all the negative charge, the electrons, in the space surrounding the nucleus, at relatively great distances.

First year chemistry or physics students knew that like charges repel, and their professors were *convinced* the Rutherford atom should fly apart. Rutherford's work also showed that most of the atom was empty space. How could the atoms that make up a solid be empty space? The solid should collapse. The experiments done by Rutherford's group were reproducible, and so had to be explained.

It seemed that the rules of classical physics that accurately predicted large-scale behavior didn't work on the atomic scale. What new rules would replace the classical ones?

Assignment 10A: Bohr's Model for Hydrogen, and Where It Led

The Rutherford model of the nuclear atom describes this tiny particle accurately but incompletely. How exactly are the electrons arranged around the nucleus? People thought that electrons circled the nucleus in orbits, like planets circle the sun. How large are these orbits? What are the positions of the electrons with respect to each other? Rutherford's model did not answer these questions.

In Assignment 10A the planetary model, as the nuclear atom is sometimes called, is enlarged and defined more specifically. The planetary model is then abandoned for the quantum mechanical model of the atom that conforms more accurately to experimental observations.

Look for these main ideas on the next page:

1) The **Bohr model** of the hydrogen atom restricts the electron to certain **quantized energy levels**. The electron can have certain definite energies, but never may it have an energy between the quantized values.

2) The **lowest quantized energy level** of the atom is called the **ground state**. All energy levels above the ground state are called **excited states**.

3) The spectrum of a gaseous atom is the result of energy released as electrons in an excited state drop to a lower energy level.

4) The **quantum mechanical model** of the atom identifies **principal energy levels and sublevels** within each principal energy level.

5) Sublevels are divided into **orbitals**, mathematically described regions of space that may be occupied by no more than two electrons.

Learning Procedures

Study

Sections 10.1–10.2, text pp. 254–262. Focus on Performance Goals 10A–10H as you study.

Answer

Questions and Problems 1–36, text pp. 282–283. Check your answers with those on text p. A.27.

Take

the skills quiz below and on the next page. Check your answers with those on SG pp. 86–87.

Assignment 10A Skills Quiz

1) Give a brief word picture of an atom according to the Bohr model.

2) Explain why the spectral lines are discontinuous when the light from an elemental gas discharge is examined in a spectroscope.

3) Are the electrons in the gas contained in an elemental gas discharge tube in the ground state or in an excited state when no electricity is passed through the tube? Explain.

4) What are the principal energy levels in an atom? Arrange them from lowest energy to highest energy.

5) List the sublevels for level $n = 4$ in order of increasing energy.

6) Sketch all the s and p orbitals for level $n = 3$.

7) How many d orbitals are there for level $n = 4$?

8) An electron orbital may be occupied by 1 or 2 electrons, but by no other number. True or false? If true, why is it true? If false, explain why it is false.

Assignment 10B: Electron Configurations and Lewis Symbols

There is a clear linkage between the shape of the periodic table, first proposed in 1869 by Mendeleev and Meyer, and the quantum mechanical model of the atom, proposed almost 60 years later. This assignment helps you to see that linkage. The main ideas of this assignment are:

1) In ground state atoms, electrons fill the lowest energy orbitals first.

2) Atoms in the same column have the same highest occupied sublevel electron configurations (**valence electrons**). The highest occupied principal energy level value increases as you go down a group.

3) Valence electrons are depicted by **Lewis symbols**, which are also called **electron dot symbols**.

4) Representative elements in the same period have the same highest occupied principal energy level, given by *n*. The highest occupied sublevel changes from *s* to *p* between groups 2A (2) and 3A (3) as you move left to right across a period.

Answer
Questions and Problems 37–64, text pp. 283–284. Check your answers with those on text p. A.27.

Take
the skills quiz below and on the next page. Check your answers with those on SG p. 87.

Learning Procedures
Study
sections 10.3–10.4, text pp. 262–270. Focus on Performance Goals 10I–10L as you study.

Assignment 10B Skills Quiz

1) Write the ground state *s* and *p* electron configuration for the highest principal energy level of the element sulfur.

2) Write complete electron configurations for the following elements:

   (a) Be _____

   (b) O _____

   (c) Al _____

   (d) K _____

   (e) Fe _____

3) Write electron configurations for the following elements, using a Group 0 (18) core:

   (a) Si _____

   (b) Na _____

   (c) Ca _____

   (d) Se _____

   (e) Ni _____

4) Draw the Lewis symbol (dot symbol) for carbon. Draw the Lewis symbol for any of the halogens.

Assignment 10C: Trends in the Periodic Table

In Assignment 10B, you saw the linkage between the periodic table and the quantum mechanical model of the atom. In this assignment, you will take another look at the periodic table, using the quantum mechanical model for the deeper insights it can provide.

Look for these big ideas in Assignment 10C:

1) **Groups** of elements in the periodic table exhibit similar behavior and are called **chemical families**.

2) The sizes of the atoms in the periodic table **increase** as you **move down** a group, but **decrease** as you move **left to right** along a period.

3) The radii of isoelectronic monatomic ions decrease as nuclear charge increases.

4) Within a chemical family, the sizes of monatomic ions increase as you go down a group in the periodic table.

5) The elements in the periodic table are classified as metals or nonmetals, based on their chemical behavior.

Learning Procedures
Study
Section 10.5, text pp. 270–277. Focus on Performance Goals 10M–10R as you study.

Answer
Questions and Problems 65–108, text pp. 284–286. Check your answers with those on text pp. A.27–28.

Take
the skills quiz below and on the next page. Check your answers with those on SG pp. 87–88.

Assignment 10C Skills Quiz

1) Explain why magnesium, Mg, calcium, Ca and barium, Ba are all members of the same chemical family.

2) List the atomic numbers of the members of the halogen family.

3) List atoms of atomic numbers 6, 7, 32 and 14 in order of increasing atomic size.

4) Arrange the ions F^-, O^{2-} and Mg^{2+} in order of *increasing* size. Explain why isoelectronic ions have different ionic radii.

5) Study the monatomic ions of elements number 11, 37 and 19. Arrange these ions in order of *decreasing* size. Explain why these ions have such different ionic radii.

6) List the atomic numbers of the nonmetals in Period 4.

7) Arrange elements of atomic numbers 32, 33 and 50 in order of *increasing* metallic character.

Answers to Chapter 10 Skills Quiz Questions

Assignment 10A

1) The Bohr model of the atom pictures electrons moving in circular orbits of fixed radii. The electrons have fixed amounts of energy and move at constant speed. Read text pp. 254–257 and check Figure 10.6, text p. 259.

2) If energy is added to any gas, atoms of that gas absorb some of the energy by having their electrons lifted from ground state to some higher quantized energy level. These excited state electrons are unstable at their high energy and fall spontaneously in one or more steps to the ground state, emitting energy equal to the energy absorbed. If the emitted energy is in the visible range, it appears as light. This process produces discrete lines in a spectroscope because electron energies are quantized, and only that specific energy between quantized energy levels may be released. Specific energy levels in the visible range result in specific colors. See the discussion on text pp. 255–256, then admire the beautiful fireworks on text p. 12 and the neon cowboy on text p. 253; both show quantum mechanics in action.

3) The electrons are in the ground state. When the electricity powering the gas tube is off, no energy is added to the atoms of gas in the tube. If atoms do not absorb energy, their electrons remain in the ground state. Check text pp. 254, 256.

4) Principal energy levels are identified by their principal quantum numbers, designated n. The principal quantum numbers are a set of whole numbers beginning with 1. Energy increases as n increases. See the summary on text p. 261.

5) In order of increasing energy, the sublevels for $n = 4$ are $4s$, $4p$, $4d$, $4f$. Read the discussion on text pp. 258–259 and the summary on text p. 261.

6) See Figure 10.7, text p. 260.

7) There are five d orbitals in each d sublevel for n greater than or equal to three. See the summary on text p. 261 and Quick Check 10.2d, on text p. 262.

8) The statement is false. An orbital may be unoccupied, that is occupied by zero electrons. An unoccupied orbital is like an unoccupied apartment. They both still exist, even if they are empty. Check text p. 260.

Assignment 10B

1) Sulfur is in Group 6A (16) in the periodic table, and so has an ns^2np^4 highest principal energy level configuration. Because sulfur is in Period 3, the electron configuration is $3s^23p^4$. Check text pp. 264–265 and Examples 10.1 and 10.5.

2) (a) Be $1s^22s^2$ (b) O $1s^22s^22p^4$ (c) Al $1s^22s^22p^63s^23p^1$
   (d) K $1s^22s^22p^63s^23p^64s^1$ (e) Fe $1s^22s^22p^63s^23p^64s^23d^6$
   Note: The $3d$ electrons may be listed ahead of the 4s electrons. Read text pp. 266–268 and check Examples 10.2–10.4, text pp. 267–268.

3) (a) Si [Ne] $3s^23p^2$ (b) Na [Ne] $3s^1$ (c) Ca [Ar]$4s^2$
   (d) Se [Ar] $4s^23d^{10}4p^4$ (e) Ni [Ar] $4s^23d^8$
   Read text pp. 266–268, especially Example 10.4, on p. 268.

4) The Lewis symbol for carbon is $\cdot \overset{\cdot}{C} \cdot$; the Lewis symbol for the halogen fluorine is $:\overset{..}{F}:$
   Check text pp. 269–270, and study Example 10.6.

Assignment 10C

1) Magnesium, calcium and barium share an ns^2 highest energy electron configuration. These two electrons are easily lost to form monatomic ions with a 2+ charge, leading to similar chemical properties in this family. Read text pp. 271–273, concentrating on the discussion of alkaline earths, p. 273.

2) The atomic numbers of the members of the halogen family are 9, 17, 35, 53. Check text p. 273 for a discussion of the halogen family.

3) 7, 6, 14, 32. Remember that "in order of increasing size" means smallest to largest, with the smallest listed first. Read the discussion on text pp. 273–275. Concentrate on the patterns on the top half of Figure 10.14, text p. 274, and see Example 10.7 on text p. 275.

4) In order of increasing size, Mg^{2+} < F^- < O^{2-}. Each of these ions has 10 electrons. The Mg^{2+} has 12 protons in its nucleus to attract 10 electrons; F^- has 9 protons to attract 10 electrons; O^{2-} has only 8 protons to attract 10 electrons. The greater the + charge in the nucleus, the stronger the electrostatic attraction between the nucleus and the electrons, and the smaller the ionic radius. Study the bottom half of Figure 10.14, text p. 274 to see this trend.

5) In order of decreasing size, Rb^+ > K^+ > Na^+. As you go down a column in the periodic table, the highest energy level electrons get further away from the nucleus. The higher the principal energy level that is occupied by valence electrons, the further away from the nucleus you find those valence electrons. Look at Figure 10.14, text p. 274 to see this trend, and read text p. 276.

6) The nonmetals in Period 4 have atomic numbers 33, 34, 35, 36. See text p. 277, particularly Figure 10.15.

7) 33, 32, 50. Read p. 277 in the text.

Sage Advice and Chapter Clues

Concentrate on the summary on text p. 261 to learn how to write electron configurations, but remember that all of these rules and summaries already exist in picture form, in the periodic table. The easiest way to master the material in Chapter 10 is to keep a "clean" periodic table next to you as you study, and use it often.

To place electrons in their proper orbitals, convince yourself that the periodic table is a grid of the elements. The position of each element on that grid tells you that element's electron configuration. The period gives you the highest value of n. For the representative elements (called either Group A or 1–2, 13–17), the group gives the highest occupied energy sublevel, and the number of electrons in that sublevel.

Some students make unneeded problems for themselves in this chapter by careless handwriting. The lower case letters s, p, d, f are symbols for subshells; upper case S, P, D and F are not. Also, the number of electrons in a subshell is given by a superscript, like $3p^2$; it is *never* given by a subscript, like $3p_2$.

Chapter 10 Sample Test

Instructions: You may use a "clean" periodic table on this test. For each multiple choice question, circle the letter of the *best* choice. Record answers to the other questions in the spaces provided.

1) The Bohr model of the atom provides all of the following *except*
   (a) an explanation of the line spectra of the elements.
   (b) a description of the electron movement in orbits around the nucleus.
   (c) evidence that electron energies are quantized.
   (d) an explanation of the differences between isotopes.
   (e) the radius of the electron orbit in a hydrogen atom.

2) Identify the *false* statement about electron energies:
   (a) Electron energies are quantized in excited states, but not quantized in the ground state.
   (b) Light spectra of the elements are experimental evidence of the quantization of electron energies.
   (c) Lines in the spectrum of an element are produced by electrons dropping from a high energy level to a lower energy level.
   (d) Energy must be absorbed to raise an electron from ground state to an excited state.
   (e) Electrons cannot possess an energy between two quantized energy levels.

3) Identify the *incorrect* statement among the following:
   (a) Except for $n = 1$, each principal energy level has three p orbitals.
   (b) There are three sublevels when $n = 3$.
   (c) Electrons in the $5d$ orbitals have higher energies than electrons in the $5p$ orbitals.
   (d) The $n = 4$ sublevels are at higher overall energies than the corresponding $n = 5$ sublevels.
   (e) A $3s$ orbital is at higher energy than a $1s$ orbital.

4) Identify the *incorrect* statement about electron orbitals:
   (a) An orbital may be occupied by no more than two electrons.
   (b) All energy sublevels contain the same number of orbitals.
   (c) For a given atom, the $3p$ orbitals are larger than the $2p$ orbitals, but smaller than the 4p orbitals.
   (d) At a given sublevel, the maximum number of d electrons is 10.
   (e) The orbital sketched at the right is a p orbital.

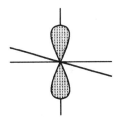

5) Using the symbolism $ns^x np^y$, where x and y are whole numbers, write the general electron configuration that is responsible for the family properties of the alkali metals.

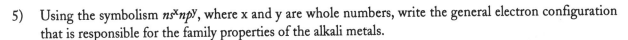

6) Write the Lewis symbol for oxygen, and then for any alkaline earth element.

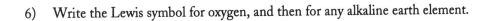

7) Write the electron configuration for an atom of chlorine.

8) Write the electron configuration for an atom of titanium, Z = 22.

9) The halogens chlorine and bromine form the insoluble silver chloride, AgCl, and silver bromide, AgBr, upon reaction with silver. These halogens exhibit these reactions with silver because:
   (a) Elemental silver has its outermost electrons in the $n = 5$ quantum level
   (b) Chlorine and bromine are in the same period of the periodic table.
   (c) Silver is a reactive metal.
   (d) Chlorine and bromine have similar electron configurations in their highest energy occupied orbitals.
   (e) Silver and these two halogens all have similar electron configurations in their highest energy occupied orbitals.

10) Which trio of atomic numbers is arranged in order of *increasing* atomic size?
    (a) 54–18–2   (b) 52–33–14   (c) 8–9–17   (d) 17–35–34   (e) 12–19–20

11) Are the ions S^{2-}, Cl^- and Ca^{2+} arranged in order of increasing or decreasing size?

12) Choose the best conclusion to the statement that begins: "The order of size of the ions in Question 12 is best explained in terms of...
    (a) the number of electrons present."
    (b) the principal quantum number of the highest occupied energy level."
    (c) the number of shielding electrons."
    (d) the magnitude of nuclear charge."
    (e) the chemical family represented by the ions."

13) Consider the monatomic ions of elements whose atomic numbers are 16, 34 and 52. The order of increasing size is:
    (a) 16–34–52   (b) 52–34–16   (c) 34–52–16   (d) 34–16–52   (e) 16–52–34

14) Consider atoms with atomic numbers 7, 11, 16, 20, 43 and 53, then
    (a) Write the atomic numbers of the metals listed above.

    (b) Write the atomic numbers of the nonmetals listed above.

Check your sample test answers with those on SG pp. 209–210.

CHAPTER 11

Chemical Bonding: The Formation of Molecules and Ionic Compounds

Assignment 11A: Ionic Bonding

A chemical bond may be loosely defined as a force that holds chemical units together. The units may be atoms, ions or molecules. Chemical bonds are formed in a system if the total energy of that system decreases after the bond is formed. There is a natural tendency towards minimization of energy. (Water flows downhill.) Ionic bonds form by the electrostatic attraction of oppositely charged particles that leads to a lower total energy when the particles are close to each other.

Ionic bonds between atoms appear to form when atoms of a metal lose one, two or three electrons to form a cation that is isoelectronic with a noble gas and atoms of a nonmetal gain one, two or three electrons to form an anion that is also isoelectronic with a noble gas. Bonds also form between two nonmetals, both of which have atoms that are one, two or even three electrons short of a noble gas electron configuration. This is accomplished through covalent bonding, in which electrons are shared. Look for these big ideas:

1) Many monatomic ions are isoelectronic with noble gas atoms, which have an octet (8) of valence electrons. The "complete octet" is a convenient memory aid.

2) An **ionic bond**, also called an **electron transfer bond**, is formed because of the electrostatic attraction between oppositely charged ions.

3) **Covalent bonds** form when two atoms achieve a noble gas octet by **sharing** one or more pairs of **electrons**.

Learning Procedures
Study
Sections 11.1–11.3, text pp. 289–295. Focus on Performance Goals 11A–11C as you study.

Answer
Questions and Problems 1–20, text pp. 303–304. Check answers with those on text p. A.28.

Take
the skills quiz on the next page. Check your answers on Study Guide (SG) p. 93.

Assignment 11A Skills Quiz

1) Write the formula of a monatomic anion and a monatomic cation that are each isoelectronic with neon.

2) Place the letter of the best choice next to each bond type:
   ionic bonding _____   covalent bonding _____
   (a) electron sharing   (b) electron transfer

Assignment 11B: All Covalent Bonds Are Not Created Equal

In Assignment 11A you learned that covalent bonds are characterized by sharing of electrons. Are the bonding electrons shared equally? If the bonding electrons are not shared equally, can we predict which of the two atoms gets the bigger share? Watch for these big ideas:

1) In a **nonpolar covalent bond**, the bonding **electrons are shared equally** by the bonded atoms. In a **polar covalent bond**, one atom attracts the shared electrons more strongly than the other.

2) The relative ability of atoms of an element to **attract electron pairs** in covalent bonds is expressed by the **electronegativity** of the element.

3) The **polarity** of a bond is estimated by the **difference in electronegativities** of the bonded atoms. The atom with the higher electronegativity is the negative end of the bond. The atom with the lower electronegativity is the positive end of the bond.

Learning Procedures

Study
Sections 11.4–11.7, text pp. 295–301. Focus on Performance Goals 11C–11F as you study.

Answer
Questions and Problems 21–42, text pp. 304–305. Check your answers with those on text p. A.29.

Take
the skills quiz below and on the next page. Check your answers with those on SG p. 93.

Assignment 11B Skills Quiz

1) Place the letter of the best choice next to each bond type:
   ionic bonding _____   covalent bonding _____   polar covalent bonding _____
   (a) unequal electron sharing   (b) electron transfer   (c) equal electron sharing

2) Tellurium has an electronegativity of 2.1; chlorine has an electronegativity of 3.0. The Te-Cl bond is best described as:
   (a) non-polar covalent   (b) polar covalent, with Te as the + end of the bond
   (c) polar covalent, with Cl as the + end of the bond
   (d) ionic, with Te as the cation and Cl as the anion.

3) Given the following electronegativity values: Li 1.0; C 2.5; Cl 3.0, consider all possible bonds between pairs of these atoms and arrange these bonds in order of increasing polarity. Classify each bond as nonpolar covalent, polar covalent or ionic. If polar covalent or ionic, show the + end and the − end of the bond.

Answers to Chapter 11 Skills Quiz Questions

Assignment 11A

1) The monatomic anions isoelectronic with neon are N^{3-}, O^{2-} and F^-; the monatomic cations isoelectronic with neon are Na^+, Mg^{2+} and Al^{3+}. Study the text pp. 289–291, especially Figure 11.1, text p. 289.

2) ionic bonding   b:electron transfer; covalent bonding   a:electron sharing   Check text pp. 293–295.

Assignment 11B

1) ionic bonding b   covalent bonding c   polar covalent bonding a
   Review Section 11.2, text pp. 292–293, and Section 11.3, text pp. 296–298.

2) Choice b is correct. An electronegativity difference of 0.9 places the bond in the polar covalent category. Because Te has a lower electronegativity than Cl, Te attracts less of the bonding electron density, and so is slightly +. Check Section 11.4, especially Examples 11.3 and 11.4, text pp. 296–298.

3) C-Cl   0.5   polar covalent: δ+ C-Cl δ−
    Li-C   1.5   polar covalent: δ+ Li-C δ−
    Li-Cl 2.0   ionic   + Li-Cl −   (note no δ here)
        Check Figure 11.6, text p. 296 and the Summary on text p. 297.

Sage Advice and Chapter Clues

Two models describe the chemical "glue" that holds the universe together. These are the ionic (electron transfer) model and the covalent (electron sharing) model. They represent extreme opposites of the bonding spectrum. As in most cases involving extreme positions, the truth is found somewhere in the middle.

Learn well the trends in electronegativity. It increases from left to right across a period and also increases from bottom to top in a group. If you remember that F is the most electronegative element, you can easily remember those trends. Electronegativity is the key to determining electrostatic attractions in bonds and in compounds. In your later chemistry courses, you will see that electronegativity is the link between many different concepts and makes them all fit together. Don't underestimate its value.

In this chapter and also in Chapter 12 you use the idea of a noble gas electron configuration, or complete octet. There is no magic in a complete octet; Section 11.7, "Exceptions to the Octet Rule" is proof that the octet rule is fallible. An ionic bond forms because of the electrostatic attractions of the ions, not because of the noble gas electron configurations those ions happen to have. Use the complete octet for what it is, a memory aid, and no more.

Chapter 11 Sample Test

Instructions: Pick the letter of the *best* choice answer to each question. You may use a "clean" periodic table.

1) Which of the following is *not* isoelectronic with a noble gas?
   (a) K^+　　　(b) F^-　　　(c) S^{2+}　　　(d) Al^{3+}　　　(e) Zn^{2+}

2) A bond formed by the transfer of an electron from one atom to another is called a(n) _____ bond:
   (a) ionic　　　(b) covalent　　　(c) polar　　　(d) nonpolar
   (e) none of the above

3) Identify the *incorrect* statement among the following:
   (a) The bond between two atoms of the same element is probably less polar than the bond between two atoms of different elements.
   (b) The distribution of electric charge in a polar bond is not symmetrical.
   (c) A bonding electron pair does not tend to be closer to either of the bonded atoms if the bond is polar.
   (d) One end of a polar bond is said to be electronegative relative to the other.

4) Given the elements and their electronegativity values: H, 2.1; C, 2.5; O, 3.5; F, 4.0; K, 0.8; Cl, 3.0, select the most polar bond below:
   (a) C-C　　　(b) H-Cl　　　(c) O-Cl　　　(d) K-F　　　(e) H-F

5) Use the electronegativity values in Question 4 to predict the polarity of a C-O bond.
   (a) nonpolar
   (b) polar covalent, with C as the + end of the bond
   (c) polar covalent, with O as the + end of the bond
   (d) ionic, with C a the cation and O as the anion.

6) The five choices below represent the difference in electronegativity between bonded atoms. Select one or more answers that describe a nonpolar or essentially nonpolar bond.
   (a) 0.0　　　(b) 1.0　　　(c) 1.5　　　(d) 2.0
   (e) none of the preceding

Check your sample test answers with those on SG p. 210.

CHAPTER 12

The Structure and Shape of Molecules

Assignment 12A: Lewis Diagrams

Chapter 11 includes a description of a covalent bond between two atoms. Chapter 12 shows how two or more of these bonds are arranged around atoms in molecules and ions.

A Lewis diagram is a representation of a molecule or ion that provides as many atoms as possible with eight valence electrons. Lewis diagrams give useful depictions of molecules or ions and also provide an accurate accounting for valence electrons. Your sole objective in this assignment is to learn to draw Lewis diagrams of polyatomic molecules or ions.

Learning Procedures
Study
Sections 12.1–12.2, text pp. 308–315. Focus on Performance Goal 12A as you study.

Answer
Questions and Problems 1–18, text p. 332. Check your answers with those on text pp. A.29, 30.

Take
the skills quiz below and on the next page. Check your answers with those on SG p. 99.

Assignment 12A Skills Quiz

Instructions: Using a "clean" periodic table, draw the Lewis diagram for each of the following substances.

1) SiH_4

2) CH_3Br

3) $AlCl_3$

4) PCl_3

5) HBr

6) BeH_2

7) NO_2^-

8) NH_2^-

Assignment 12B: Molecular Geometry and Polarity

Lewis diagrams give us some idea about the structure of molecules, but they can be misleading. A Lewis diagram is simply an accounting device to keep track of the valence electrons. It often represents a three dimensional molecule on a two dimensional piece of paper or computer screen. The physical and chemical properties of compounds are largely the result of the two and three dimensional shapes (geometries) and polarities of their molecules.

Look for these big ideas:

1) A bond angle is the angle between any two bonds formed by the same atom.

2) **Electron pair geometry** describes the arrangement of two, three or four pairs of electrons, either shared or unshared, around a central atom.

3) **Molecular geometry** describes the arrangement of two, three or four atoms around a central atom to which they are all bonded.

4) Electron pair and molecular geometries are said to be linear, bent (angular), trigonal planar, tetrahedral or pyramidal (trigonal pyramid).

5) Electron pair, and therefore molecular, geometry may be predicted by the **valence shell electron pair repulsion (VSEPR)** principle.

6) The contribution to molecular geometry of a multiple bond is the same as if it were a single bond.

7) The polarity of a molecule may be predicted from its shape and the polarity of its bonds. Some molecules with polar bonds are nonpolar because of their molecular geometries.

Learning Procedures
Study
Sections 12.3–12.6, text pp. 316–323, top. Focus on Performance Goals 12B–12D as you study.

Answer
Questions and Problems 19–44, text pp. 332–333. Check your answers with those on text p. A.31.

Take
the skills quiz on the next page. Check your answers with those on SG pp. 96–100.

Assignment 12B Skills Quiz

Instructions: Predict the electron pair and molecular geometry for those substances whose Lewis diagrams you drew in the Assignment 12A Skills Quiz. State whether each substance is polar or nonpolar. If a substance is polar, indicate the electropositive and electronegative regions.

1) SiH_4

2) CH_3Br

3) $AlCl_3$

4) PCl_3

5) HBr

6) BeH_2

7) NO_2^-

8) NH_2^-

Assignment 12C: Some Organic Compounds

Over 11,000,000 compounds have been identified since 1965 in the chemistry laboratories of the world. Of these, about 95% are classed as organic compounds, compounds based upon the carbon atom and carbon-carbon bonds. This section gives a quick overview of some organic compounds made from carbon, hydrogen, and oxygen.

The main ideas in this optional section are:

1) Because carbon atoms bond to each other forming long chains, there is immense variety in the number and shapes of carbon compounds.

2) **Hydrocarbons** are made of **carbon and hydrogen**. The **alkanes** are a hydrocarbon family with **all single bonds**.

3) **Alkenes** are hydrocarbons with carbon-carbon double bonds; **alkynes** are hydrocarbons with carbon-carbon triple bonds.

4) Compounds with the same molecular formula but different molecular structures are called **isomers**.

5) If a hydrogen atom in a hydrocarbon (for example, CH_4) is replaced by a hydroxyl group, -OH, the resulting product is an **alcohol**, CH_3-OH. An alcohol may also be viewed as a water molecule (H-O-H) in which *one* hydrogen atom is replaced by a hydrocarbon group (CH_3-OH.)

6) An **ether** may be viewed as a water molecule (H-O-H) in which *both* hydrogen atoms are replaced by hydrocarbon groups (CH_3-O-CH_3.) Alcohols and ethers with the same number of carbon atoms are isomers.

7) If a hydrogen atom in a hydrocarbon is replaced by a carboxyl group, -COOH, the resulting product is a **carboxylic acid**. The most common carboxylic acid is acetic acid, written as $HC_2H_3O_2$ or CH_3COOH.

Learning Procedures
Study
Section 12.7, text pp. 325–330. Focus on Performance Goals 12E–12G as you study.

Answer
Questions and Problems 45–60, text pp. 333–334. Check your answers with those on text p. A.31.

Take
the skills quiz below and on the next page. Check your answers with those on SG p. 100.

Assignment 12C Skills Quiz

Instructions: Select the letter of the *best* choice for 1; for 2 and 3 write the answer in the space provided.

1) Which of these is considered to be organic?
    (a) CO   (b) CO_3^{2-} (c) $HC_2H_3O_2$ (d) CN^-

2) Write in your own words a definition of the word isomer. Give examples of isomers by drawing two Lewis structures that have the formula C_5H_{12}.

3) From the line formulas given below, select the hydrocarbons, the alcohols, the ethers and the carboxylic acids. State which formulas are isomers.
    $CH_3CH_2CH_2CH_3$ $CH_3CH_2CH_2$-O-CH_3 $CH_3CH_2CH_2$-COOH $CH_3CHOHCH_2CH_3$

Answers to Chapter 12 Skills Quiz Questions

Assignment 12A

For Problems 1-5, read text pp. 308–309; see Example 12.1, text pp. 308–309.

1) H—Si—H with H above and H below (all single bonds to Si)

2) H—C—H with :Br: above and H below

3) Al with :Cl: above, :Cl: lower-left, :Cl: lower-right

4) :Cl—P—Cl: with :Cl: above

5) H—Br: (with lone pairs)

6) H—Be—H   For 6, read text pp. 311–312.

7) $[\ddot{O}=\dot{N}-\ddot{O}:]^{-}$ 8) $[H-\ddot{N}-H]^{-}$ For 7 and 8, see text pp. 310–315, Examples 12.2–12.7.

Assignment 12B

For all of the problems in this assignment, read about electron pair and molecular geometries on text pp. 316–322, especially Table 12.1, text p. 318. The line references are for Table 12.1. Read text pp. 323–325 on polarity for all these problems. Specific references follow each problem.

1) Both the electron pair and molecular geometry of SiH_4 are tetrahedral (line 3). The SiH_4 molecule is nonpolar; study Figure 12.7, text p. 324.

2) Both the electron pair and molecular geometry of CH_3Br are tetrahedral (line 3). The CH_3Br molecule is polar, with C electropositive and Br electronegative. Read text pp. 323–324, and Figure 12.7.

3) Both the electron pair and molecular geometry of $AlCl_3$ are trigonal planar (line 2). This molecule is nonpolar and also does not obey the octet rule. See Example 12.9, text p. 322 for shape and Example 12.11, text p. 322 for polarity.

4) The electron pair geometry of PCl_3 is tetrahedral, but the molecular geometry is pyramidal (line 4). The PCl_3 molecule is polar, with P electropositive and Cl electronegative. Read Example 12.11, text pp. 324–325 on polarity.

5) The molecular geometry of HBr is linear; the HBr molecule is polar, with H electropositive and Br electronegative. Read p. 264 in the text.

6) Both the electron pair and molecular geometry of BeH_2 are linear; this molecule is nonpolar. Read line 1 for geometry information, and read the second paragraph in Section 12.6, text p. 323 for polarity information.

7) The electron pair geometry of NO_2^- is trigonal planar, but the molecular geometry is bent. See Section 12.5, text pp. 322–323. The NO_2^- ion is polar, with N electropositive and O electronegative.

8) The electron pair geometry of NH_2^- is tetrahedral, but the molecular geometry is bent (line 5). The NH_2^- ion is polar, with H electropositive and N electronegative. Compare Problem 12.40, text p. 333.

Assignment 12C

1) The correct answer is c. Read text p. 325.
2) Isomers are substances that have the same molecular formula, but different order of attachment of their atoms. The three possible isomers of C_5H_{12} are shown below. Check text pp. 326–327.

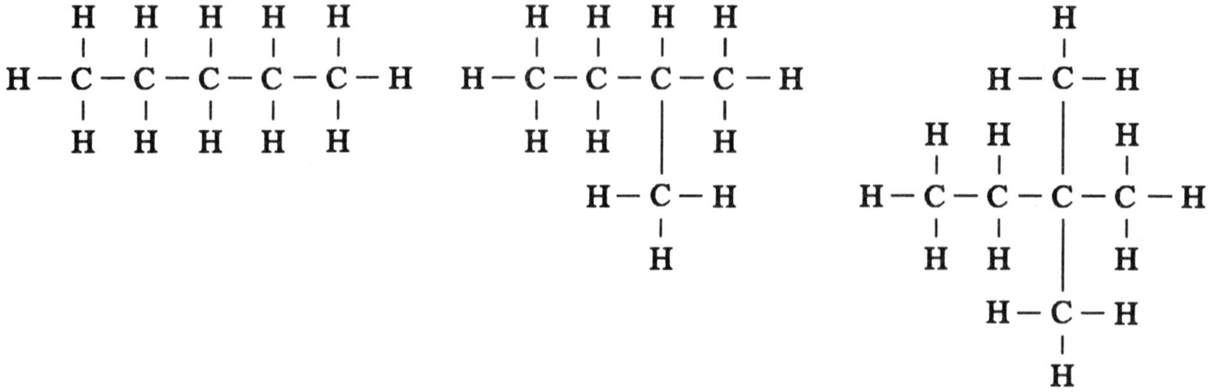

```
  H H H H H              H H H H                        H
  | | | | |              | | | |                        |
H-C-C-C-C-C-H        H-C-C-C-C-H                    H-C-H
  | | | | |              | | | |                        |
  H H H H H              H H | H              H  H      | H
                             |                |  |      | |
                           H-C-H          H-C-C-C-C-H
                             |                |  |   |
                             H                H  H   |
                                                  H-C-H
                                                    |
                                                    H
```

3) The formulas depict, from left to right, a hydrocarbon, an ether, a carboxylic acid, an alcohol. The ether and alcohol, both with formula $C_4H_{10}O$, are isomers. Read text pp. 327, bottom to 330, top, and Problems 55–56 on text p. 333.

Sage Advice and Chapter Clues

You can frequently draw more than one Lewis diagram that satisfies the octet rule for a given substance. Sometimes these additional diagrams are correct for other real substances, and sometimes they are not. You cannot tell if a compound exists just because it has an acceptable Lewis diagram; you also need laboratory evidence. However, there are some generalizations that guide you to diagrams that are most apt to be correct. They are on text pp. 310–311. Two other generalizations also come in handy: 1) Because H forms only one bond, you can't put a hydrogen atom in the middle of a Lewis diagram; 2) in polyatomic species, the least electronegative atom usually goes in the middle.

When you must predict geometries, keep in mind that there are two types of geometry, electron pair geometry and molecular geometry. Electron pair geometry describes the orientation of the electron pairs around a central atom; molecular geometry describes the orientation of atoms bonded to a central atom. The two geometries are the same only if there are no lone pair electrons around the central atom. These geometries are summarized in Table 12.1, page 318 in the text.

Hydrocarbons come in so many types, it might be easier to remember these types by remembering an example of each type that you already know. The alkanes have all single bonds; you may have used liquid propane or butane to run camp stoves. Cycloalkanes have a carbon chain in a closed ring, or cycle. Cyclopropane is an operating room anesthetic. Many plastics are made from alkenes, compounds with a carbon-carbon double bond. Some names for these plastics are polyethylene (food wrap, plastic milk bottles) or polypropylene (ropes, indoor-outdoor carpets.) The most common alkyne, a compound with a carbon-carbon triple bond, unfortunately has the misleading name acetylene, used in welding and cutting torches. The aromatic hydrocarbons are based on benzene, which has an alternating single, double bond structure in a ring. Styrene is a $H_2C=CH-$ group connected to a benzene ring. It is made into plastics such as polystyrene and Styrofoam.

You can remember the single bonded oxygen organic compounds as water molecules with a hydrocarbon group replacing hydrogen:

    water   H–O–H
    alcohol   R–O–H
    ether   R–O–R

where R is an hydrocarbon group (missing one H) that completes the Lewis structure.

Try this to remember carboxylic acids. The hydroxyl group, -OH, is a contraction of hydrogen and oxygen. (This also explains the hydrox*ide* anion, OH^-.) The carboxyl group, -COOH, is a contraction of carbon, hydrogen and oxygen. These names come down from the 19th century, and so are not systematic. Memory is needed here.

Chapter 12 Sample Test

Instructions: Draw Lewis diagrams for each substance in Questions 1–8. You may use a "clean" periodic table.

1) OF_2

2) NH_3

3) SeO_4^{2-} (For Se, Z=34)

4) NO_3^-

5) H_3PO_4

6) C_3H_4

7) C_3H_7F

8) $C_3H_6O_2$

For Questions 9–12, predict the electron pair and molecular geometry of the substances from Questions 1–4.

	Electron Pair Geometry		Molecular Geometry
9)	_____	(OF_2)	_____
10)	_____	(NH_3)	_____
11)	_____	(SeO_4^{2-})	_____
12)	_____	(NO_3^-)	_____

For Questions 13–14, draw Lewis diagrams for the substance indicated, and state whether it is nonpolar or polar. If polar, indicate the electropositive and the electronegative regions.

13) BHF_2

14) $BeBr_2$

15) Write the Lewis structures of all the organic compounds having the formula C_4H_8. Identify these as alkanes, alkenes, alkynes, cycloalkanes or aromatic hydrocarbons.

16) Write the Lewis structures of all the organic compounds having the formula C_2H_6O. Identify the alcohols and ethers.

Check your sample test answers with those on SG pp. 210–211.

CHAPTER

13

The Ideal Gas Law and Gas Stoichiometry

Assignment 13A: Old and New Two Variable Gas Laws

In Chapter 4 you learned that there are four measurable properties of a gas: pressure, volume, temperature and quantity. The gas laws in Chapter 4 kept quantity constant. Because you now know about the mole, we can add quantity to our measurable gas properties.

The first three big ideas given below are review items from Chapter 4. If you need to review these items, do so *before* going on. Chapter 13 starts with item 4.

1) **Gay-Lussac's Law** states that at constant volume, the pressure of a fixed quantity of a gas is **directly proportional** to the absolute temperature, $P \propto T$.

2) **Charles' Law** states that at constant pressure, the volume of a fixed quantity of a gas is **directly proportional** to the absolute temperature, $V \propto T$.

3) **Boyle's Law** states that at constant temperature, the volume of a fixed quantity of a gas is **inversely proportional** to its pressure, $V \propto 1/P$. (Remember that in an inverse proportionality, one variable increases as the other decreases.)

4) **Avogadro's Law** states that **equal volumes** of two gases at the same temperature and pressure contain the **same number of molecules**, $V \propto n$.

Learning Procedures

Study
Sections 13.1–13.3, text pp. 336–339. If needed, review Sections 4.5–4.8, text pp. 96–106.

Answer
Questions and Problems 1–2, text p. 356. Check your answers with those on text p. A.31.

Take
the skills quiz below and on the next page. Check your answers with those on SG p. 107.

Assignment 13A Skills Quiz

1) A gas sample has a pressure of 516 torr at 24°C; what is the pressure of this sample if the temperature is changed to 91°C with no volume change?

2) The volume of a gas is 5.56 L at 819 torr. If the pressure is lowered to 622 torr with no temperature change what is the new gas volume?

3) A gas occupies 1.29 L at 82°C and 617 torr. What volume would this gas occupy at 95°C and 954 torr?

4) Container A holds nitrogen at a given temperature and pressure. Container B, equal in volume to container A, holds ethane, C_2H_6, at the same temperature and pressure. In which container is the mass of gas the greatest? Explain in terms of Avogadro's Law.

Assignment 13B: The Ideal Gas Equation and its Applications. Molar Volume and Molar Mass

The proportional relationships between pressure, volume, quantity and temperature of a gas make it possible to combine them in a single equation. This is known as the ideal gas equation, and it may be used for any gas that behaves in accord with the model of an ideal gas.

There are five Performance Goals for this assignment that all come together in the ideal gas equation. Look for this single, recurring theme through the following ideas:

1) The ideal gas equation is **PV = nRT**.

2) Two values of the **Universal gas constant, R**, are 0.0821 L · atm/mol · K = 62.4 L · torr/mol · K.

3) Given all the values in the ideal gas equation but one, the remaining value may be calculated.

4) The density of a gas is directly proportional to its molar mass. Either molar mass or density can be calculated from the other using the ideal gas equation.

5) **Molar volume** is the number of liters occupied by one mole of a gas. The molar volume of any ideal gas at STP is 22.4 L/mol. This quantity is useful in calculations involving moles, mass, volume, density and molar mass of a gas measured at STP.

Learning Procedures
Study
Sections 13.3–13.4, text pp. 338–344, top. Focus on Performance Goals 13A–13E as you study.

Answer
Questions and Problems 3–44, text pp. 356–357. Check your answers with those on text pp. A. 31–33.

Take
the skills quiz on the next page. Check your answers with those on SG pp. 107–108.

Assignment 13B Skills Quiz

1) Express the temperature of a gas in terms of the ideal gas equation.

2) What is the volume of 14.0 g of nitrogen gas at 20°C and 935 torr pressure?

3) Calculate the density of carbon dioxide gas at 16°C and a pressure of 698 torr.

4) The density of a gas at STP is 1.88 grams per liter. Calculate the molar mass of this unknown gas.

5) Find the molar mass of an unknown gas if 5.58 grams occupy 5.23 L at 0.483 atmospheres and 80°C.

6) Calculate the mass of carbon monoxide in 7.80 L at STP.

7) What is the molar volume of ammonia at 31°C and 756 torr?

Assignment 13C: Gas Stoichiometry, Dalton's Law of Partial Pressures

Back in Chapter 9 you were promised that if you learned how to use the stoichiometry pattern in solving problems involving mass, you would also know how to solve problems with other quantity units such as energy. If you did the learning, you can cash in again on the promise. If you forgot, review text pp. 223–224.

In Assignment 13B you learned how to convert between moles of a gas and volume at stated temperature and pressure. Converting between measured quantity and moles is the first step and the third step in the stoichiometry pattern. With gases, the quantity is usually measured in volume units.

Here are the major ideas in Assignment 13C:

1) The volume of a gas at specified temperature and pressure may be used for quantity in solving stoichiometry problems.

2) If given and wanted quantities in a stoichiometry problem are gas volumes at different temperatures and pressures, convert the volume of the given gas to what it would be at the desired temperature and pressure, then complete the problem by a volume to volume conversion using the coefficients in the chemical equation.

3) The **partial pressure** of a gas in a gaseous mixture is the pressure that **gas alone** would exert in the same volume at the same temperature.

4) The **total pressure** of a gas mixture is the **sum of the partial pressures** of all gases in that mixture.

Learning Procedures

Study
Sections 13.5–13.7, text pp. 344–355. Focus on Performance Goals 13F–13H as you study.

Answer
Questions and Problems 45–62, text pp. 357–358. Check your answers with those on text p. A.33.

Take
the skills quiz below and on the next page. Check your answers with those on SG pp. 108–109.

Assignment 13D Skills Quiz

1) Calculate the volume of NH_3, measured at 22°C and 745 torr, that can be produced by 7.52 grams of nitrogen in the reaction $N_2 + 3 H_2 \rightarrow 2 NH_3$.

2) What volume of oxygen, measured at 165 torr and 28°C, is required to react with 50.0 L of sulfur dioxide, measured at 85°C and 790 torr, to produce sulfur trioxide by the reaction $2 SO_2 + O_2 \rightarrow 2 SO_3$?

3) The partial pressure of carbon dioxide in a mixture of two gases is 0.86 atm.; the partial pressure of carbon monoxide in that mixture is 1.21 atm. What is the pressure of the mixture?

4) Find the partial pressure of oxygen collected over water at 28.4°C at a total pressure of 761 torr. The vapor pressure of water at this temperature is 29 torr.

Answers to Chapter 13 Skills Quiz Questions

Assignment 13A

1)

	Volume	Temperature	Pressure	Amount
Initial Value (1)	constant	24°C = 297 K	516 torr	constant
Final Value (2)	constant	91°C = 364 K	P_2	constant

$516 \text{ torr} \times \dfrac{(273 + 91) \text{ K}}{(273 + 24) \text{ K}} = 632 \text{ torr}$ See text pp. 96–98, especially Example 4.3 on 97–98.

2)

	Volume	Temperature	Pressure	Amount
Initial Value (1)	5.56 L	constant	819 torr	constant
Final Value (2)	V_2	constant	622 torr	constant

$5.56 \text{ L} \times \dfrac{819 \text{ torr}}{622 \text{ torr}} = 7.32 \text{ L}$ See text pp. 100–103, particularly Examples 4.5–4.6.

3)

	Volume	Temperature	Pressure	Amount
Initial Value (1)	1.29 L	82°C = 355 K	954 torr	constant
Final Value (2)	V_2	95°C = 368 K	617 torr	constant

$1.29 \text{ L} \times \dfrac{(273 + 95) \text{ K}}{(273 + 82) \text{ K}} \times \dfrac{617 \text{ torr}}{954 \text{ torr}} = 0.865 \text{ L}$ See Example 4.7, text p. 104; ponder the picture on p. 87.

4) The mass of ethane is greater than the mass of nitrogen. Avogadro's Law states that equal volumes of different gases at the same temperature and pressure contain the same number of molecules. The containers therefore hold the same number of moles of ethane and nitrogen. The molar mass of ethane (30 g/mole) is larger than the molar mass of nitrogen (28 g/mole), so the mass of ethane must also be larger. Read text pp. 337–338 and see Figures 13.1, 13.2.

Assignment 13B

1) $PV = nRT$; $T = \dfrac{PV}{nR}$

Study text pp. 338–339. If your algebra is a bit rusty, please check out "SOLVING AN EQUATION FOR AN UNKNOWN QUALITY," on text p. A.7. Do it now!!! You'll be glad you did....

Note: Each of the following answers starts with Equation 13.3 and then derives whatever version of Equation 13.3 or Equation 13.6 is needed. The moral of this story: It all goes back to a single equation, PV = nRT.

2) GIVEN: 14.0 g N_2; 28.0 g/mol; 935 torr; 62.4 L·torr/mol·K; 20°C (293 K)
   WANTED: L N_2 EQUATIONS: V = nRT/P; n = m/MM; V = mRT/MM·P

$$V = \frac{m\,RT}{MM \cdot P} = 14.0 \text{ g } N_2 \times \frac{1 \text{ mol } N_2}{28.0 \text{ g } N_2} \times \frac{62.4 \text{ L·torr}}{\text{mol·K}} \times \frac{(273 + 20) \text{ K}}{935 \text{ torr}} = 9.78 \text{ L}$$

Check Example 13.1, text pp. 339–340. If you forgot that nitrogen is diatomic, shame on you. Go back and review Section 6.2, text pp. 137–138.

3) GIVEN: 16°C (289 K); 698 torr; 62.4 L·torr/mol·K; 44.0 g/mol CO_2
   WANTED: Density, m(ass)/V, units of g/L   EQUATION: m/V = (MM)P/RT

$$D = \frac{m}{V} = \frac{MM \cdot P}{RT} = \frac{44.0 \text{ g}}{1 \text{ mol}} \times \frac{698 \text{ torr}}{289 \text{ K}} \times \frac{1 \text{ mol·K}}{62.4 \text{ L·torr}} = 1.70 \text{ g/L}$$ Study Example 13.13, text p. 341.

4) GIVEN: gas density, 1.88 g/L; 273 K; 1.00 atm;   WANTED: molar mass, g/mol

$$\text{EQUATION: } MM = \frac{m\,RT}{VP} = \frac{1.88 \text{ g}}{1.00 \text{ L}} \times \frac{0.0821 \text{ L·atm}}{\text{mol·K}} \times \frac{273 \text{ K}}{1.00 \text{ atm}} = 42.1 \text{ g/mol}$$

Shrewdly note that the last two terms above multiply out to 22.4 L/mole, the molar volume at STP; this problem could then also be solved as follows:

$$\frac{1.88 \text{ g}}{1 \text{ L}} \times \frac{22.4 \text{ L}}{\text{mole}} = 42.1 \text{ g/mole}$$

See Quick Check 13.4, text p. 342.
Don't forget, *this only works for a gas at STP!*

5) GIVEN: 5.58 g; 5.23 L; 80°C (353 K); 0.483 atm; 0.0821 L·atm/mol·K
   WANTED: molar mass, g/mol;   EQUATIONS: n = PV/RT ; n = m/MM; MM = m/n

$$MM = \frac{mRT}{PV} = \frac{5.58 \text{ g}}{5.23 \text{ L}} \times \frac{(273 + 80) \text{ K}}{0.483 \text{ atm}} \times \frac{0.0821 \text{ L·atm}}{\text{mol·K}} = 64.0 \text{ g/mole}$$ Example 13.4 text pp. 341–342.

6) GIVEN: 7.80 L CO; 1.00 atm; 273 K; 28.0 g/mol CO   WANTED: g CO
   EQUATIONS: m/MM = n; m = n(MM); n = PV/RT; m = (MM)PV/RT

$$m = \frac{28.0 \text{ g CO}}{1 \text{ mol CO}} \times \frac{1.00 \text{ atm}}{273 \text{ K}} \times \frac{0.0821 \text{ L·atm}}{\text{mole·K}} \times 7.80 \text{ L} = 9.75 \text{ g CO}$$ See Problem 17, text p. 357.

7) GIVEN: 31°C (304 K); 756 torr; 62.4 L·torr/mol·K
   WANTED: MV (L/mol)   EQUATIONS: MV = V/n; PV = nRT; V/n = RT/P

$$\frac{V}{n} = \frac{RT}{P} = \frac{62.4 \text{ L·torr}}{\text{mol·K}} \times \frac{304 \text{ K}}{756 \text{ torr}} = 25.1 \text{ L/mol}$$ See Quick Check 13.5, text p. 343.

Assignment 13C

1) GIVEN: 22°C (295 K); 745 torr; 62.4 L·torr/mol·K;   WANTED: MV, V/n

EQUATION: $MV = \dfrac{V}{n} = \dfrac{RT}{P} = \dfrac{62.4 \text{ L·torr}}{\text{mol·K}} \times \dfrac{295 \text{ K}}{745 \text{ torr}} = 24.7 \text{ L/mol}$ We have the molar volume, now on to the stoichiometry!

GIVEN: 7.52 g N_2 WANTED: L NH_3 PATH: g $N_2 \to$ mol $N_2 \to$ mol $NH_3 \to$ L NH_3

FACTORS: 28.0 g/mol N_2; 2 mol NH_3/mol N_2; 24.7 L/mol N_2

$7.52 \text{ g } N_2 \times \dfrac{1 \text{ mol } N_2}{28.0 \text{ g } N_2} \times \dfrac{2 \text{ mol } NH_3}{1 \text{ mol } N_2} \times \dfrac{24.7 \text{ L } NH_3}{\text{mol } NH_3} = 13.3 \text{ L } NH_3$

OR...

GIVEN: 7.52 g N_2 WANTED: mol NH_3 PATH: g $N_2 \to$ mol $N_2 \to$ mol NH_3

FACTORS: 28.0 g /mol N_2; 2 mol NH_3/mol N_2

$7.52 \text{ g } N_2 \times \dfrac{1 \text{ mol } N_2}{28.0 \text{ g } N_2} \times \dfrac{2 \text{ mol } NH_3}{1 \text{ mol } N_2} = 0.537 \text{ mol } NH_3$ Now on to the gas law stuff!

GIVEN: 0.537 mol NH_3; 22°C (295 K); 745 torr, 62.4 L·torr/mol·K   WANTED: L NH_3

EQUATION: $V = \dfrac{nRT}{P} = 0.537 \text{ mol } NH_3 \times \dfrac{62.4 \text{ L·torr}}{\text{mol·K}} \times \dfrac{(273 + 22) \text{ K}}{745 \text{ torr}} = 13.3 \text{ L } NH_3$

Study text pp. 344–351, especially Example 13.9, text pp. 346–347.

3)

	Volume	Temperature	Pressure	Amount
Initial Value (1)	50.0 L	85°C = 358 K	790 torr	constant
Final Value (2)	V_2	28°C = 301 K	165 torr	constant

$50.0 \text{ L } SO_2 \times \dfrac{(273 + 28) \text{ K}}{(273 + 85) \text{ K}} \times \dfrac{790 \text{ torr}}{165 \text{ torr}} \times \dfrac{1 \text{ L } O_2}{2 \text{ L } SO_2} = 101 \text{ L } O_2$ See Example 13.12, text pp. 350–351.

4) 1.21 + 0.86 = 2.07 atm   5) 761 − 29 = 732 torr   See text, pp. 351–353.

Sage Advice and Chapter Clues

When using the ideal gas equation, solve the equation algebraically for the wanted quantity, and then substitute the given values, with *complete* units. Don't solve the problem yet. If the unit cancellation does not give the wanted quantity, the solution is incorrect. Write the complete units on the gas constant, R, too. You will then be certain you are using the correct value for R, and that you have converted temperature from °C to K. *Now* get out your calculator and solve the problem.

As you solve stoichiometry problems involving gases, be sure to recognize that the pattern is identical to that used in mass stoichiometry. The stoichiometry pattern, first presented on page 225 of the text and repeated on page 344, is applied in both cases. The only difference is in the quantity unit being converted to moles, or vice versa. In one case it is in grams, and in the other it is gas volume at specified temperature and pressure.

Chapter 13 Sample Test

Instructions: For 1, choose the letter of the *best* choice. Answer the remaining questions in the space provided.

1) One of two identical containers holds oxygen, and the other chlorine. Both gases exert a pressure of 1.19 atm. at 21°C. Which statement is *incorrect*?
   (a) The number of molecules of oxygen is the same as the number of molecules of chlorine.
   (b) The mass of the oxygen is equal to the mass of the chlorine.
   (c) The number of moles of oxygen is equal to the number of moles of chlorine.
   (d) The number of oxygen atoms is equal to the number of chlorine atoms.

2) Express the gas constant, R, in terms of the ideal gas equation.

3) What is the density (g/L) of ammonia at STP?

4) Find the molar mass of a gas if 0.460 L, measured at 819 torr and 22°C, has a mass of 0.369 grams.

5) What is the molar volume of fluorine gas at -17°C and 1.03 atm?

6) The molar volume of hydrogen bromide gas at 14°C and 772 torr is 23.2 L/mol; how many mol of gas are in a 1.25 L vessel at these conditions?

7) Calculate the grams of zinc that must react to produce 148 mL of hydrogen gas at 767 torr and 24°C by the reaction $Zn + 2\ HCl \rightarrow H_2 + ZnCl_2$.

8) What volume of oxygen, measured at 0.891 atm. and 18°C, is needed to burn completely 4.18 L of butane, C_4H_{10}, measured at 1.34 atm. and 38°C?
The gas phase reaction is: $2\ C_4H_{10} + 13\ O_2 \rightarrow 8\ CO_2 + 10\ H_2O$.

9) In a mixture of gases, the partial pressure of carbon dioxide is 241 torr, of carbon monoxide, 337 torr, and of nitrogen 413 torr. Calculate the total pressure of this three-gas mixture.

10) A mixture of nitrogen, oxygen and water has a total pressure of 2174 torr. If the partial pressure of the water is 151 torr, and the partial pressure of nitrogen is 1162 torr, what is the partial pressure of oxygen?

Check your sample test answers with those on SG pp. 211–212.

CHAPTER 14

Liquids and Solids

Assignment 14A: Intermolecular Forces in the Liquid State

The ideal gas equation from Chapter 13 actually deals with the spaces between gas molecules, not the molecules themselves. That is why one equation can show the relationship between the measurable properties of many different gases. Gas molecules are very far apart, and so the attractive forces between them in a gas are negligible. In liquids, however, molecules are very close to each other, so intermolecular forces are strong. No single equation describes the measurable properties of liquids; many of these properties are determined by intermolecular attractions in the liquid.

Here are the main ideas to learn in Assignment 14A:

1) Certain **physical properties of liquids** such as vapor pressure, boiling point, heat of vaporization, viscosity and surface tension can be explained in terms of **intermolecular attractions**.

2) **Dipole forces** give an electrostatic attraction between polar molecules.

3) **Dispersion forces**, also called **London forces**, are the **comparatively weak** intermolecular attractions between nonpolar molecules. London forces are the result of temporary dipoles caused by shifting electron density in molecules. London forces vary directly with surface area and may be large if the molecules are large.

4) **Exceptionally strong** dipole-like forces called **hydrogen bonds** arise between molecules that have hydrogen atoms bonded to a highly electronegative atom. This atom, usually nitrogen, oxygen or fluorine, must have at least one unshared electron pair.

Learning Procedures
Study
Sections 14.1–14.2, text pp. 362–369. Focus on Performance Goals 14A–14E as you study.

Answer
Questions and Problems 1–36, text pp. 390–391. Check your answers with those on text p. A.34.

Take
the skills quiz on the next page. Check your answers with those on Study Guide (SG) p. 118.

Assignment 14A Skills Quiz

Instructions: Select the letter of the *best* choice answer.

1) Intermolecular forces are weaker in gases than in liquids because:
   - (a) liquids are less polar than gases.
   - (b) molecules are far apart in liquids.
   - (c) molecules are far apart in gases.
   - (d) liquids are less dense than gases.

2) Equal volumes of liquids A and B, at the same temperature, are poured through identical funnels. Each funnel has a long, narrow stem. Liquid A requires more time to flow through the funnel than liquid B. The intermolecular attractions in liquid A are _____ the intermolecular attractions in liquid B.
   - (a) stronger than
   - (b) equal to
   - (c) weaker than
   - (d) not able to be compared to

3) Select the *incorrect* statement about intermolecular forces.
   - (a) Dispersion forces exist in all molecules.
   - (b) Hydrogen bonding is a stronger intermolecular force than dipole force.
   - (c) Dispersion forces are weaker than dipole forces.
   - (d) Hydrogen bonding requires a weak hydrogen-hydrogen bond between two molecules.

4) Liquid CCl_4 is held together by:
   - (a) dipole forces only
   - (b) dispersion forces only
   - (c) both dipole forces and dispersion forces
   - (d) neither dipole forces nor dispersion forces

5) Based upon the generalizations in this assignment, you would expect SiH_3Cl to have _____ boiling point than CH_3Cl. (electronegativities: H = 2.1; C = 2.5; Si = 1.8; Cl = 3.0)
   - (a) a lower
   - (b) the same
   - (c) a higher

Assignment 14B: Liquid-Vapor Equilibrium and Boiling

In this assignment we examine in detail vapor pressure and boiling point. Both of these physical properties are associated with the change from the liquid to the vapor state. Intermolecular attractions must be overcome when a liquid changes to a gas. The ease with which this is done determines both the pressure exerted by the vapor at a temperature, and the temperature at which boiling occurs.

The main ideas you should identify in this assignment are:

1) **Equilibrium** is defined as the condition in which the **rates of opposing changes are equal**. In a liquid-vapor equilibrium, the rate of evaporation is equal to the rate of condensation.

2) The partial pressure exerted by a vapor in equilibrium with a liquid is the **equilibrium vapor pressure** at the existing temperature.

3) Equilibrium vapor pressure increases with temperature. At higher temperatures, a larger frac-

tion of the liquid sample has enough energy to evaporate.

4) The **boiling point** of a liquid is the temperature at which the **vapor pressure** of that liquid is **equal to** or slightly greater than the **surrounding pressure**. Normal boiling point is the boiling point at one atmosphere of pressure.

5) Water, a molecule necessary for life, breaks almost all the rules for predicting physical properties of liquids due to its extremely strong hydrogen bonding.

Learning Procedures
Study
Sections 14.3–14.5, text pp. 369–374. Focus on Performance Goals 14F–14H as you study.

Answer
Questions and Problems 37–62, text pp. 391–393. Check answers with those on text p. A.34.

Take
the skills quiz below. Check your answers with those on SG p. 118.

Assignment 14B Skills Quiz

Instructions: Choose the letter of the *best* choice answer.

1) A highly volatile liquid, CH_2Cl_2, is placed in a container. The container is then sealed. Some, but not all of the liquid disappears. The vapor pressure of the CH_2Cl_2 in the container
    (a) is lower than the equilibrium vapor pressure.
    (b) is higher than the equilibrium vapor pressure.
    (c) is the same as the equilibrium vapor pressure.
    (d) can't be compared to equilibrium vapor pressure without more details.

2) A properly stated boiling point includes both a boiling point temperature and the surrounding pressure because
    (a) vapor pressure of a liquid is not constant at a given temperature.
    (b) boiling points increase in a vacuum.
    (c) boiling points decrease as vapor pressure increases.
    (d) boiling points decrease as surrounding pressure decreases.

3) The dew point is the temperature at which atmospheric water vapor is in equilibrium with liquid water. At temperatures below the dew point, dew or fog (both forms of liquid water) occurs because
    (a) water vapor molecules move at slower speeds than water liquid molecules.
    (b) below the dew point, the rate of condensation is faster than the rate of evaporation.
    (c) below the dew point, the rate of condensation is slower than the rate of evaporation.
    (d) as temperature decreases, the vapor pressure of water increases due to hydrogen bonding.

4) Water, H_2O, boils at 100°C; hydrogen sulfide, H_2S, boils at -86°C. This is best explained by
    (a) stronger dispersion forces between H_2O molecules than between H_2S molecules.
    (b) stronger hydrogen bonds between H_2O molecules than between H_2S molecules.
    (c) H_2S having greater polarity than H_2O.
    (d) none of these.

Assignment 14C: Solids

According to the kinetic molecular theory, particles of a solid do not move past each other, but simply "shake" in their assigned places in a crystal. The beauty of crystals has fascinated people for ages. Look at Figures 14.15–14.18, pp. 376–377 in the text for examples of this beauty.

Amazingly, all crystals fall into only seven geometric forms, and this assignment is limited to the identification of a few crystal forms and types of solids. Look for the following ideas in this short assignment:

1) In a **crystalline solid**, the particles, either ions or molecules, are arranged in a **definite geometric form** that runs throughout a perfect crystal, or throughout each part of an imperfect crystal. An **amorphous solid** has **no long range order** in the arrangement of particles.

2) **Crystalline solids** may be classified as **ionic, molecular, network** or **metallic**. Each crystal type is associated with characteristic physical properties.

Learning Procedures
Study
Sections 14.6–14.7, text pp. 374–377. Focus on Performance Goals 14I–14J as you study.

Answer
Questions and Problems 63–66 text p. 393. Check your answers with those on text p. A.34.

Take
the skills quiz below. Check your answers with those on SG p. 118.

Assignment 14C Skills Quiz

Instructions: Select the *best* choice answer for each question.

1) Both window glass and quartz are made up largely of silicon dioxide. Window glass softens over a wide temperature range, while quartz melts at 1610°C. Window glass is most likely a(n)_____ solid while quartz is most likely a(n)_____ solid.
   (a) crystalline; amorphous
   (b) amorphous; crystalline
   (c) opaque; crystalline
   (d) amorphous; molecular

2) Substance A melts at 2810°C and is insoluble in water. Substance A conducts electricity when melted, but not when solid. Substance A is most likely a(n)_____ crystal.
   (a) metallic   (b) network   (c) molecular   (d) ionic

Assignment 14D: Energy, Temperature Changes, Physical Changes

To change the temperature of something you must warm it or cool it. To change a liquid to a gas, heat must be added; to change a liquid to a solid, heat must be removed. How much heat must be added or removed to accomplish a given change of temperature or state depends on two things, how much matter is being changed, and certain physical properties of the substance.

The main ideas in this assignment are:

1) **Heat of vaporization** (or condensation) is the

energy transferred when one gram of a substance changes between the liquid and gaseous states.

2) **Heat of fusion** (or solidification) is the energy transferred when one gram of a substance changes between the liquid and solid states.

3) **Specific heat** is the amount of energy needed to change the temperature of one gram of a substance one degree Celsius.

4) Heat transfer problems involve either three or four factors. You can solve them either using dimensional analysis or algebraically.

5) For a pure substance, a graph of temperature versus energy added to or removed from that substance expresses temperature change in the sloped regions and change of state in the horizontal regions. Heat flow from one point to another on the graph is the sum of the heat flows for each step in the process. Figure 14.20 on text p. 383 illustrates this.

Learning Procedures

Study
Sections 14.8–14.10, pp. 378–386 in the text. Focus on Performance Goals 14K–14O as you study.

Answer
Questions and Problems 67–106, text pp. 393–395. Check answers with those on text pp. A.34–35.

Take
the skills quiz below and on the next page. Check your answers with those on SG pp. 118–119

Assignment 14D Skills Quiz

Instructions: Solve each problem in the space provided.

1) The heat of vaporization of ammonia is 1.37 kJ/g. What heat flow is necessary to boil 26.3 grams of ammonia, if the ammonia is already at its boiling point?

2) What mass of copper metal is present if 2.20 kJ are needed to melt the metal, already at its melting point? The heat of fusion of copper metal is 205 J/g.

3) The specific heat of acetone is 2.13 J/g · °C. How much heat flow is needed to raise 34.3 g of acetone from 25°C to 45°C?

4) Calculate the specific heat of a liquid if 808 joules are required to warm 28.0 grams of this liquid from 18.4°C to 32.7°C.

5) The graph to the right presents a temperature versus energy plot for a pure substance. Identify the physical states of the substance at each numbered point on the graph, and note the region on the graph corresponding to the heat of fusion.

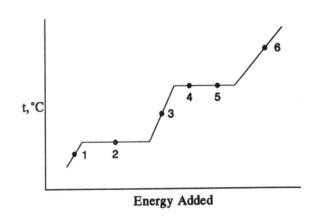

6) What is the total heat flow if 114 grams of ice, $H_2O(s)$, initially at -81°C warms to water at 39°C? The specific heat of ice is 2.1 J/g · °C; the specific heat of water is 4.18 J/g · °C. The heat of fusion of water is 335 J/g, and the melting point of water is 0°C.

Answers to Chapter 14 Skills Quiz Questions

Assignment 14A

1) Choice c is correct. Read text p. 362.

2) Choice a is correct. Study text p. 364, top.

3) Choice d is correct. Read text pp. 365–369, especially Figures 14.6 and 14.7, text pp. 367–368.

4) Choice b is correct. Although each C-Cl bond is CCl_4 is polar, the tetrahedral molecular geometry cancels these dipoles out. See Question 5.

5) Choice c is correct. SiH_3Cl boils at +1.9°C; CH_3Cl boils at -24°C. SiH_3Cl has a higher molar mass than CH_3Cl, and the electronegativity of Si is lower than that of C, so the silicon compound also has more polar bonds, and therefore higher polarity. The higher the polarity and molar mass, the higher the boiling point. For both Questions 4 and 5, read text pp. 365–369, and Figure 14.7 on page 368.

Assignment 14B

1) Choice c is correct. Read text pp. 369–371, especially Figure 14.9, p. 370.

2) Choice d is correct. See text pp. 372–373. Note that Figure 14.10, text p. 371, also gives boiling points of three liquids as the surrounding pressure changes. When the vapor pressure of a liquid equals the surrounding pressure, boiling occurs. For example, H_2O boils at 60°C at about 150 torr pressure. Go back and convince yourself that fact comes from Figure 14.10.

3) Choice b is correct. Read text pp. 371–372 and see Quick Check 14.5.

4) Choice b is correct. Check pp. 373–374 in the text, pausing to ponder the penguins.

Assignment 14C

1) Choice b is correct. Read text pp. 374–375, then check Figure 14.17 on text p. 376.

2) Choice d is correct. The conductivity is the key here. Check text pp. 375–377, Quick Check 14.8 and Table 14.3 on p. 377.

Assignment 14D

1) GIVEN: 1.37 kJ/g, 26.3 g   WANTED: heat flow in kJ   EQUATION: $Q = m\Delta H_{vap}$
   $Q = (26.3 \text{ g})(1.37 \text{ kJ/g}) = 36.0 \text{ kJ}$ Study text pp. 378–380, and Examples 14.1–14.3.

2) GIVEN: 2.20 kJ, 205 J/g   WANTED: mass in g
   EQUATION: $Q = m\Delta H_{fus}$ $m = \dfrac{Q}{\Delta H_{fus}} = \dfrac{2200 \text{ J}}{205 \text{ J/g}} = 10.7 \text{ g}$ See Example 14.3, p. 380, and Problem 79, p. 394.

3) GIVEN: 2.13 J/g · °C, 34.3 g, T_i = 25°C, T_f = 45°C WANTED: Q
EQUATION: Q = mcΔT = (34.3 g)(2.13 J/g · °C)(45°C − 25°C) = 1,461 J = 1.5 kJ
Read text pp. 380–382, and see Example 14.4, text pp. 381–382.

4) GIVEN: 808 J, 28.0 g, T_i = 18.4°C, T_f = 32.7°C WANTED: c EQUATION: Q = mcΔT
$$Q = mc\Delta T \quad c = \frac{Q}{m\Delta T} = \frac{808 \text{ J}}{(28.0 \text{ g})(14.3°C)} = 2.02 \text{ J/g} \cdot °C$$
Review Example 14.5, text p. 382. Note the sign convention on ΔT.

5) At point 1, only solid exists; at point 2 there is a mixture of solid and liquid. At point 3 there is liquid only; at points 4 and 5, there is a mixture of liquid and vapor. Point 6 contains only vapor. The horizontal region near point 2 is associated with the heat of fusion. Study the text pp. 382–385, paying careful attention to Figure 14.20 on page 383.

6) GIVEN: 114 g H_2O(s), T_i = −81°C, T_f = 39°C, freezing point = 0°C
Specific heats: 2.1 J/g · °C (s), 4.18 J/g · °C, (ℓ), ΔH_{fus} = 335 J/g
WANTED: total heat flow Q_{total} EQUATIONS: Q = mcΔT, Q = mΔH_{fus}

Q_{total} = Q_{ice} + $Q_{\Delta fus}$ + Q_{water}
Q_{ice} = (114 g)(2.1 J/g · °C)[(0°C − (−81°C)]   = 19391 J   = 19 kJ
$Q_{\Delta fus}$ = (114 g)(335 J/g)   = 38190 J   = 38 kJ
Q_{water} = (114 g)(4.18 J/g·°C)(39°C − 0°C)   = 18584 J   = 19 kJ
See Figure 14.20, text p. 383, and Example 14.6, pp. 384–385. Q_{total} = 76 kJ

Sage Advice and Chapter Clues

There are three types of intermolecular attractions. If all other things are equal, their order of increasing strength is generally dispersion forces, dipole forces and hydrogen bonds. All of these forces are much weaker than chemical bonds. Strong intermolecular attractions lead to an increase in all of the physical properties mentioned in this chapter except vapor pressure. Remember that vapor pressure results from the gas produced by evaporation, and that strong intermolecular attractions tend to prevent the evaporation of a liquid. The amount of vapor present, and hence the vapor pressure, are low when little liquid can evaporate.

Ionic crystals and network crystals are held together by chemical bonds, and so they exhibit high melting points. Molecular crystals are held together by much weaker intermolecular attractions, so they have low melting points. The melting points of metallic crystals vary greatly depending on the size of the metal atoms, and the number of "freely moving" valence electrons.

Be certain that you understand hydrogen bonding. The term "hydrogen bond" is unfortunately misleading. The hydrogen bond is not a chemical bond like those you studied in Chapters 10 and 11. Chemical bonds are *intra*molecular (within one molecule;) hydrogen bonds are usually *inter*molecular (between two or more molecules.)

The first three assignments in this chapter dealt with intermolecular attractions; the fourth assignment dealt with the energy changes needed to overcome these attractions and change temperature or change a solid to a liquid, or a liquid to a gas. Most of the information you need to remember about these energy changes is given in Figure 14.20, truly a picture worth a thousand words. The sloped regions of this graph contain only one physical state; the horizontal regions contain a mixture of physical states, either solid and liquid, or liquid and gas. The heat terms associated with the horizontal regions are ΔH_{fusion} or $\Delta H_{vaporization}$. Because in these regions heat can be added or withdrawn without a change in temperature, you might also see these terms called "latent," or hidden heats.

Problems involving specific heat, heat flow, ΔT or mass are best solved using Equation 14.3 (p. 378) or 14.5 (p. 380) at phase change points or Equation 14.6 (p. 381) for heating or cooling. Always check that the units cancel properly in the algebraic setup.

What if you have trouble remembering $Q = mc\Delta T$?

Think of your utility bills last winter, when you paid for some Q to heat the place in which you live. How could you make Q, and those bills, smaller? You could: 1) lower the thermostat, to change ΔT between inside and outside; 2) insulate, to change the "specific heat," c, of your place; 3) if all else fails, you close off rooms or move to a smaller place, so there's less "mass" m to heat. $Q = mc\Delta T$. It's a useful equation.

Chapter 14 Sample Test

Instructions: You may use a "clean" periodic table. Select the *best* answer for each multiple choice question and circle the letter of that choice. Solve each numeric question in the space provided.

1) The primary reason intermolecular forces are stronger in liquids than in gases is that
    (a) liquids are cooler than gases.
    (b) molecules are closer to each other in liquids.
    (c) liquids weigh more than gases.
    (d) the liquid state is between the gaseous state and the solid state.

2) The bulb of a medicine dropper is depressed and the tip is immersed into liquid A. When the bulb is released the dropper fills quickly. After cleaning the dropper the identical procedure is followed with liquid B, but the filling is slower, more sluggish. From these observations...
    (a) it is reasonable to predict that intermolecular attractions are stronger in A than in B.
    (b) it is reasonable to predict that intermolecular attractions are about the same in A and in B.
    (c) it is reasonable to predict that intermolecular attractions are stronger in B than in A.
    (d) no reasonable prediction can be made about the relative strengths of intermolecular attractions in A and B.

3) Select from the following the statement about intermolecular forces that is *incorrect*.
    (a) Attractions between molecules without hydrogen bonding are generally weaker than attractions between hydrogen bonded molecules of about the same size.
    (b) Dispersion forces exist only between nonpolar molecules.
    (c) Dipole forces exist only between polar molecules.
    (d) Hydrogen bonding would not be evident in molecules in which all the hydrogen is covalently bonded to sulfur and no other element.

4) What type(s) of intermolecular forces would you expect in $CHF_3(\ell)$?
   (a) dispersion only
   (b) dipole and dispersion
   (c) dispersion and hydrogen bonding
   (d) dipole only

5) Based on the generalizations developed in this chapter, list the halogen oxides below in order of *increasing* boiling point.
   (a) OF_2, Cl_2O, Br_2O
   (b) Cl_2O, OF_2, Br_2O
   (c) Cl_2O, Br_2O, OF_2
   (d) Br_2O, OF_2, Cl_2O

6) Closed system A consists of liquid CCl_4 in equilibrium with its own vapor at 20°C. System B is identical to system A, except that the equilibrium temperature is 40°C. Identify the *incorrect* statement among the following:
   (a) The vapor pressure of A is less than that of B.
   (b) The evaporation rate of A is less than that of B.
   (c) The condensation rate of A is less than that of B.
   (d) The vapor concentration of A is greater than that of B.

7) Acetone, a highly volatile liquid, is placed in a container, and the container is sealed. The liquid disappears. The vapor pressure of the acetone in the container
   (a) is lower than the equilibrium vapor pressure.
   (b) is higher than the equilibrium vapor pressure.
   (c) is the same as the equilibrium vapor pressure.
   (d) cannot be compared to the equilibrium vapor pressure without additional information.

8) The normal boiling points of the two major components of air, nitrogen and oxygen, are -196°C and -183°C, respectively. Liquid air at -200°C is in a closed cylinder with gaseous helium (b. pt. -269°C) at 760 torr above the surface of the liquid. If the pressure of the helium is slowly released,
   (a) nitrogen will begin to boil before oxygen.
   (b) nitrogen and oxygen will begin to boil at the same time.
   (c) oxygen will begin to boil before nitrogen.
   (d) you cannot predict which liquid will boil first.

9) Silly Putty™ is a slightly elastic solid that bounces like a rubber ball when dropped. This elasticity suggests that Silly Putty™ is a(n) _____ solid.
   (a) molecular crystalline
   (b) ionic crystalline
   (c) amorphous
   (d) network crystalline

10) Germane melts at -165°C, is insoluble in water, and does not conduct electricity when melted. Germane most likely forms a(n) _____ crystal.
    (a) ionic
    (b) metallic
    (c) network
    (d) molecular

11) Calculate the heat of vaporization in kJ/g of sodium metal if 10.0 kJ energy is needed to boil 2.35 grams of sodium metal that is already at the boiling point.

12) How much heat, in kJ, is needed to melt 123 grams of Ni(s), already at its melting point, if the heat of fusion for Ni(s) is 310 J/g?

13) The graph to the right presents a temperature versus energy plot for a pure substance. Identify the region(s) on the graph where only liquid exists, and the heat of vaporization.

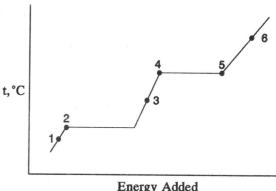

14) Calculate the heat absorbed by 3636 grams of lubricating oil in a car engine as it warms from garage temperature (16°C) to operating temperature (110°C); the specific heat of the oil is 3.6 J/g · °C.

15) A 65.0 gram sample loses 1.92 kJ when cooling from 114.6°C to 31.2°C. Calculate the specific heat of the sample.

16) What is the total heat flow if 44 grams of $H_2O(\ell)$, initially at 82°C is cooled to $H_2O(s)$ at -23°C? The specific heat of $H_2O(\ell)$ is 4.18 J/g · °C; the specific heat of $H_2O(s)$ is 2.1 J/g · °C. The heat of fusion of water is 335 J/g, and the freezing point of water is 0.0°C.

Check your sample test answers with those on SG pp. 213.

CHAPTER 15

Solutions

Assignment 15A: Characteristics of Solutions

To a chemist, the word **solution** means a homogeneous mixture, as well as the explanation to a problem. This assignment is an introduction to solution characteristics and terminology. Learn the big ideas:

1) The **major component** of a solution is called the **solvent**; the minor components are called the **solutes**. Every solution has **one solvent**, and one or more solutes.

2) Unlike the composition of a pure substance, the percentage composition of solutions may vary.

3) Two or more substances will mix to form a solution only if the intermolecular attractions in each separate substance are similar.

Learning Procedures
Study
Sections 15.1–15.4, text pp. 398–403. Make an outline and focus on the subheadings of your outline as you study.

Answer
Questions and Problems 1–34, text pp. 426–428 (Don't overlook Table 15.2, text p. 428.) Check your answers with those on text p. A.36.

Take
the skills quiz below and on the next page. Check your answers with those on Study Guide (SG) p. 131.

Assignment 15A Skills Quiz

Instructions: Answer each question in the space provided.

1) Champagne is a superb example of a solution, for champagne is mainly water(ℓ) with added ethyl alcohol(ℓ), CO_2(g) and sugar(s). Name the solvent and solute(s) in this delightful solution.

2) (a) How can you make a dilute solution from a concentrated solution?

(b) How can you make a concentrated solution from a dilute solution?

3) You can dissolve 14 g of solute Y in 100 g water to form a saturated solution. On the scale below, label the saturated, supersaturated and unsaturated regions for solute Y in water.

```
|————————————————————————————————————————|
0                10                20                30
              grams of Y in 100 grams of water
```

4) Gasoline and water do not dissolve in each other. Are they miscible or immiscible liquids?

5) Making candy often calls for dissolving a large amount of sugar in water to form a saturated sugar solution. Both the sugar crystal and the liquid water have dipole forces and hydrogen bonding as intermolecular attractions. Describe this dissolving process in words, then explain how you could speed up this process.

6) Diesel cars are usually equipped with water traps, to remove water from diesel fuel. Diesel fuel is a hydrocarbon mixture with average formula of $C_{20}H_{42}$. Predict if water and diesel fuel will mix, and state your reasons for your prediction.

7) The partial pressure of CO_2 in the atmosphere is about 0.3 torr. If this partial pressure increases to 0.6 torr, how would bottlers change the amount of CO_2 in their carbonated beverages, to maintain constant "fizziness?"

Assignment 15B: Solution Concentrations

Solutions are mixtures, not pure substances, so the composition of a solution is not fixed by a molecular formula. We need ways to specify quantitatively the composition of a solution. The textbook identifies four ways that are covered in this assignment. Check with your instructor if you are studying all four topics, or only the first two. Look for these new ideas:

1) The **percentage concentration** of any component in a solution is the mass of that component in a sample divided by the total mass of that sample, all multiplied by 100.

2) **Molarity** is moles of solute in one liter of solution. Units of molarity are **moles per liter**, mol/L; the symbol for molarity is **M**. The **number of mol** in a known volume of a solution of known molarity is given by $V \times M$, where the volume, V, is given in liters.

3) **Molality** is the moles of solute dissolved in one kilogram of solvent. The symbol for molality is m (lower case to differentiate it from the upper case M for molarity.)

4) **One equivalent of an acid** is defined as that amount of acid that yields **one mole of hydrogen ion in a specific reaction**. One equivalent of a **base** is that amount of base that **reacts with one equivalent of an acid.**

5) The **equivalent mass** of a substance is the number of **grams** of the substance **per equivalent**.

6) **Normality** is the number of equivalents of solute in one liter of solution. Units of normality are **equivalents per liter**, or **eq/L**; the symbol for normality is **N**. The **number of equivalents** in a known volume of a solution of known normality is given by $V \times N$, where the volume, V, is given in liters.

Learning Procedures

Study
Sections 15.5–15.9, text pp. 403–414. Make an outline and focus on the subheadings of your outline as you study.

Answer
Questions and Problems 35–84, text pp. 428–429. Check your answers with those on text pp. A.36, 37.

Take
the skills quiz on the next *two* pages. Check your answers with those on SG pp. 131–132.

Assignment 15B Skills Quiz

Instructions: Solve each problem in the space provided.

1) A water solution contains 153 g Na_3PO_4 and 415 g H_2O. What is the percentage concentration of Na_3PO_4 in this solution?

2) What mass of water and calcium chloride, $CaCl_2$, would you need to make 1562 g of a 5.03% $CaCl_2$ solution?

3) What is the molarity of the solution made by dissolving 12.9 g solid sodium bromide in water and diluting to a final volume of 5.00×10^2 mL?

4) How many grams of K_2SO_4 do you need to make 196 mL of a 0.317 M potassium sulfate solution?

5) If 3.14 g of CHI_3 are dissolved in 23.52 g of solvent, what is the molality of the CHI_3 in this solution?

6) Explain what an equivalent of base is.

7) State the number of equivalents of acid per mole H_3PO_4 and the equivalent mass of H_3PO_4 in the following reaction:

$$2\ H_3PO_4(aq) + 3\ Ca(OH)_2(s) \rightarrow Ca_3(PO_4)_2(s) + 6\ HOH(\ell)$$

8) How many grams of KOH are needed to make 855 mL of 0.716 N KOH?

Assignment 15C: Solution Dilution and Stoichiometry

Molarity is the most widely used of the four solution concentrations introduced in Assignment 15B. In this assignment you will learn two important calculation skills involving molarity: 1) Given concentrated solutions, add water to make dilute solutions; 2) Use molarity in stoichiometry problems such as titrations.

1) Because dilution problems involve adding only water to a solute, the number of moles of solute remains unchanged in the dilution process.

2) Combining big idea #2 of Assignment 15B with item 1, above, we obtain the equation $M_c \times V_c = M_d \times V_d$, in which $_c$ stands for "concentrated" and $_d$ stands for "dilute."

3) Number of moles obtained from molarity and volume data can be used in the three step stoichiometry pattern learned earlier.

4) **Titration** is the **controlled and measured addition** of one solution into another. Figure 15.6, text p. 417 shows this process.

Learning Procedures
Study
Sections 15.10–15.12, text pp. 414–419. Make an outline and focus on the subheadings of your outline as you study.

Answer
Questions and Problems 85–110, text pp. 429–431.
Check your answers with those on text pp. A.37–38.

Take
the skills quiz below. Check your answers with those on SG p. 133.

Assignment 15C Skills Quiz

Instructions: Solve each problem in the space provided.

1) Concentrated acetic acid is 17.5 M; vinegar is 0.87 M acetic acid. How much concentrated acetic acid do you need to make 32 L of vinegar?

2) Find the number of mL of 0.103 M KCl required to precipitate as lead chloride all the lead in a solution that contains 0.293 g lead nitrate.

3) Find the molarity of a solution of sodium hydroxide if 21.4 mL are required to react with 20.0 mL 0.101 M hydrochloric acid in a titration experiment.

4) Sodium carbonate is used as a primary standard for acids. In a titration, 24.46 mL of hydrochloric acid are needed to titrate 0.259 g sodium carbonate. Calculate the molarity of the hydrochloric acid. The equation is: $Na_2CO_3(aq) + 2\ HCl(aq) \rightarrow 2\ NaCl(aq) + CO_2(g) + H_2O(\ell)$.

5) What is the volume of $CO_2(g)$, measured at STP, produced in the titration from Question 4, above?

Assignment 15D: Normality Titrations, Colligative Properties (optional)

With the increasing use of SI units, normality as a concentration unit has become a controversial topic among chemists. The SI system does not acknowledge normality as an acceptable concentration unit. Nevertheless, normality and normality titrations are so convenient for both academic and industrial laboratories that normality will die only after a long hard battle, and perhaps never.

Because it is a mixture, a solution has physical properties that vary with its concentration. The temperature at which a solution freezes or boils depends on the molality of solute in that solution.

The major ideas in this assignment are:

1) In a reaction, the **numbers of equivalents of acid and base** that react with each other are **equal**. This idea of equal numbers of equivalents is the basis of the normality system.

2) Combine item 1 above and big idea #6 from Assignment 15B (V × N = number of equivalents) to obtain for an acid-base titration, $V_{acid} \times N_{acid} = V_{base} \times N_{base}$.

3) The properties of a solution that depend only on the number of solute particles present, without regard to their identity, are called **colligative properties**.

4) Freezing point depression and boiling point elevation are colligative properties that are directly proportional to the molal concentration (Assignment 15B, big idea #3) of any solute. These proportionality constants are called the **molal freezing point constant** and the **molal boiling point constant**, respectively. The values of these constants depend only on the chemical identity of the solvent in the solution.

5) For freezing point depression, $\Delta T_f = K_f m$; for boiling point elevation, $\Delta T_b = K_b m$.

6) **Freezing point depression** experiments are often used to **determine molar mass**.

Learning Procedures
Study
Sections 15.13–15.14, text pp. 419–423. Make an outline and focus on the subheadings of your outline as you study.

Answer
Questions and Problems 111–136, text pp. 431–432. Check your answers with those on text p. A.38.

Take
the skills quiz below and on the next page. Check your answers with those on SG p. 134.

Assignment 15D Skills Quiz

Instructions: Solve each problem in the space provided.

1) Find the mL of 0.107 N NaOH needed to react with 24.2 mL of 0.153 N H_2SO_4 in the reaction $H_2SO_4(aq) + 2\,NaOH(aq) \rightarrow Na_2SO_4(aq) + 2\,HOH(\ell)$.

2) Exactly 22.73 mL of sulfuric acid is needed to titrate 0.504 g of sodium carbonate in the reaction
$$H_2SO_4(aq) + Na_2CO_3(aq) \rightarrow CO_2(g) + Na_2SO_4(aq) + H_2O(\ell).$$
What is the normality of the sulfuric acid in this reaction?

3) Find the normality of a solution of sodium hydroxide if 23.7 mL are required to react with 25.0 mL of 0.136 N phosphoric acid in the reaction
$$3\ NaOH(aq) + H_3PO_4(aq) \rightarrow Na_3PO_4(aq) + 3\ HOH(\ell).$$

4) Ethylene glycol, $C_2H_6O_2$, is automotive antifreeze. What will be the freezing point of a solution of 83.1 g ethylene glycol in 112 g of water? The molal freezing point constant for water is 1.86°C/m.

5) The molal freezing point constant for water is 1.86 °C/m. If 3.52 g of an unknown are dissolved in 20.79 g water, the resultant solution freezes at -1.97°C, as measured with a thermometer that gives the freezing point of pure water as 0.00°C. What is the molar mass of this unknown?

Answers to Chapter 15 Skills Quiz Questions

Assignment 15A

1) The solvent is water, the substance present in greatest amount; the solutes are ethyl alcohol, CO_2 and sugar. Check text p. 399, Solute & Solvent.

2) You can make a dilute solution from a concentrated solution by either adding more solvent, or by removing some of the solute. You can make a concentrated solution from a dilute solution by either adding more solute, or by removing some solvent, perhaps by evaporation. Study p. 399 in the text.

3) Read pp. 399–400 in the text.

```
                           saturated at 14 g
                                  |
_____
0<-----<---unsaturated----->-----10-------->-- 14 --<----<-----<--- 20 ---<-----supersaturated----->30
                   grams of Y in 100 grams of water
```

4) Water and gasoline are immiscible. Read text p. 400.

For Questions 5 and 6, study text pp. 400–402.

5) The sugar molecules start by being locked in a crystal; they can vibrate, but not move past each other. When dropped into water, the sugar molecules at the outside of the crystal are attracted by the water. If a sugar molecule becomes hydrated (dissolved), the attractions to the water solvent are the same types as those that held it in the crystal. To speed up the dissolving process you could heat the mixture, stir the mixture, and use more finely divided sugar crystals (confectioner's, or powdered sugar rather than granulated sugar).

6) Water molecules are small and polar, held together by dipole forces and hydrogen bonding. Diesel fuel has large nonpolar molecules held together by dispersion forces. These very different molecules have different intermolecular attractions, so they do not mix.

7) Bottlers would, in theory, need to increase the amount of CO_2 dissolved in their beverages. The solubility of a gas dissolved in a liquid increases as the partial pressure of that gas above the liquid increases. If its solubility is higher, more gas dissolves before saturation is reached, and it fizzes out. Read text pp. 402–403, particularly Figure 15.3.

Assignment 15B

For 1 and 2, read text pp. 381-383; see the examples listed for each question.

1) GIVEN: 153 g Na_2SO_4, 415 g H_2O WANTED: % Na_2SO_4

EQUATION: $\% = \dfrac{\text{g solute}}{\text{g solute + g solvent}} \times 100 = \dfrac{153 \text{g}}{153 \text{ g} + 415 \text{ g}} \times 100 = 26.9\%$ See Example 15.2, text p. 404.

2) GIVEN: 1562 g solution, 5.03 g $CaCl_2$/100 g solution   WANTED: g $CaCl_2$
   PATH: g solution → g $CaCl_2$

   $$1562 \text{ g solution} \times \frac{5.03 \text{ g } CaCl_2}{100 \text{ g solution}} = 78.6 \text{ g } CaCl_2$$

   1562 g solution = g H_2O + g $CaCl_2$ = 1562 g − 79 g $CaCl_2$ = 1483 g H_2O
   Study Example 15.3, text pp. 404–405.

3) GIVEN: 12.9 g NaBr, 5.00×10^2 mL (0.500 L), 102.9 g NaBr/mol NaBr
   WANTED: M   PATH: g NaBr → mol NaBr → M_{NaBr}
   EQUATION: M = mol/L

   $$12.9 \text{ g NaBr} \times \frac{1 \text{ mol NaBr}}{102.9 \text{ g NaBr}} = 0.125 \text{ mol NaBr} \qquad \frac{0.125 \text{ mol NaBr}}{0.500 \text{ L}} = 0.251 \text{ M}$$

   Using the "concentration ratio" method on text pp. 406–407, we can write:

   $$\frac{12.9 \text{ g NaBr} \times \frac{1 \text{ mol NaBr}}{102.9 \text{ g NaBr}}}{0.500 \text{ L}} = \frac{12.9 \text{ g NaBr}}{0.500 \text{ L}} \times \frac{1 \text{ mol NaBr}}{102.9 \text{ g NaBr}} = 0.251 \text{ M NaBr}$$

   Note that the 0.125 mol NaBr is actually 0.125*4* mol. If this intermediate answer is kept in your calculator, you get the above answer, rounded from 0.250*8* M. Read text pp. 405–408, Example 15.5.

4) GIVEN: 196 mL (0.196 L), 0.317 mol K_2SO_4/L, 174.3 g K_2SO_4/mol K_2SO_4 WANTED: g K_2SO_4
   PATH: L → mol K_2SO_4 → g K_2SO_4

   $$0.196 \text{ L} \times \frac{0.317 \text{ mol } K_2SO_4}{\text{L}} \times \frac{174.3 \text{ g } K_2SO_4}{1 \text{ mol } K_2SO_4} = 10.8 \text{ g } K_2SO_4$$

   See Example 15.4, text p. 405.

5) GIVEN: 3.14 g CHI_3, 23.52 g solvent (0.02352 kg), 393.7 g CHI_3/mol CHI_3 WANTED: m
   PATH: $\dfrac{(\text{g } CHI_3 \to \text{mol } CHI_3)}{\text{kg solvent}}$

   $$\frac{3.14 \text{ g } CHI_3}{0.02352 \text{ kg solvent}} \times \frac{1 \text{ mol } CHI_3}{393.7 \text{ g } CHI_3} = 0.339 \text{ m } CHI_3$$

   Read text pp. 408–409, study Example 15.8, text pp. 408–409.

6) An equivalent of base is that amount of base that reacts with one mole of hydrogen ions. See text p. 410.

7) In this reaction, there are 3 eq acid/mol H_3PO_4, from the equation.

   $$\frac{98.0 \text{ g } H_3PO_4}{1 \text{ mol } H_3PO_4} \times \frac{1 \text{ mol } H_3PO_4}{3 \text{ eq acid}} = 32.7 \text{ g } H_3PO_4/\text{eq acid}$$

   Read text pp. 409–412, Examples 15.10–15.11.

8) Because KOH can neutralize only one proton, no equation is needed as there can be only 1 eq/mol KOH.
   GIVEN: 855 mL (0.855 L), 56.1 g KOH/mol KOH, 1 eq KOH/mol KOH, 0.716 eq/L KOH
   WANTED: g KOH   PATH: L KOH → eq KOH → mol KOH → g KOH

   $$0.855 \text{ L KOH} \times \frac{0.716 \text{ eq KOH}}{\text{L}} \times \frac{1 \text{ mol KOH}}{1 \text{ eq KOH}} \times \frac{56.1 \text{ g KOH}}{\text{mol KOH}} = 34.3 \text{ g KOH}$$

   See Example 15.16, text pp. 412–414.

Assignment 15C

1) GIVEN: $M_c = 17.5$ mol/L_c, $M_d = 0.87$ mol/L_d, $V_d = 32$ L WANTED: V_c

EQUATION: $V_c = \dfrac{V_d \times M_d}{M_c} = \dfrac{32\ L_d \times 0.87\ mol/L_d}{17.5\ mol/L_c} = 1.6\ L = 1.6 \times 10^2\ mL$ See Examples 15.17 and 15.18, text p. 415.

For 2–4, read text pp. 415–419 and read the examples listed for each question.

2) $Pb(NO_3)_2(aq) + 2\ KCl(aq) \rightarrow PbCl_2(s) + 2\ KNO_3(aq)$ Write the equation!

GIVEN: 0.103 mol KCl/L, 0.293 g $Pb(NO_3)_2$, 331.2 g $Pb(NO_3)_2$/mol $Pb(NO_3)_2$, 2 mol KCl/mol $Pb(NO_3)_2$

WANTED: mL KCl   PATH: g $Pb(NO_3)_2$ → mol $Pb(NO_3)_2$ → mol KCl → L KCl → mL KCl

$0.293\ g\ Pb(NO_3)_2 \times \dfrac{1\ mol\ Pb(NO_3)_2}{331.2\ g\ Pb(NO_3)_2} \times \dfrac{2\ mol\ KCl}{1\ mol\ Pb(NO_3)_2} \times \dfrac{1\ L\ KCl}{0.103\ mol\ KCl} \times \dfrac{1000\ mL}{L} = 17.2\ mL\ KCl$

See Example 15.19, text p. 416.

3) $NaOH(aq) + HCl(aq) \rightarrow NaCl(aq) + HOH(\ell)$ Once more, write the equation!

GIVEN: 0.101 mol HCl/L, 20.0 ml HCl (0.0200 L), 1 mol HCl/mol NaOH, 21.4 mL NaOH (0.0214 L)

WANTED: M NaOH   PATH: L HCl → mol HCl → mol NaOH

EQUATION: M = mol/L

$0.0200\ L\ HCl \times \dfrac{0.101\ mol\ HCl}{L\ HCl} \times \dfrac{1\ mol\ NaOH}{1\ mol\ HCl} \times \dfrac{1}{0.0214\ L\ NaOH} = 0.0944\ M\ NaOH$

Check Examples 15.22–15.24, text pp. 418–419, particularly 15.24.

4) GIVEN: 0.259 g Na_2CO_3, 106.0 g Na_2CO_3/mol, 2 mol HCl/mol Na_2CO_3, 24.46 mL HCl (0.02446 L)

WANTED: M HCl   PATH: g Na_2CO_3 → mol Na_2CO_3 → mol HCl → M_{HCl}

EQUATION: M = mol/L

$0.259\ g\ Na_2CO_3 \times \dfrac{1\ mol\ Na_2CO_3}{106.0\ g\ Na_2CO_3} \times \dfrac{2\ mol\ HCl}{1\ mol\ Na_2CO_3} \times \dfrac{1}{0.02446\ L\ HCl} = 0.200\ M\ HCl$

See Examples 15.22–15.23, text pp. 418–419.

5) GIVEN: 0.259 g Na_2CO_3, 106 g Na_2CO_3/mol, 1 mol CO_2/mol Na_2CO_3, WANTED: L CO_2

PATH: g Na_2CO_3 → mol Na_2CO_3 → mol CO_2 → L CO_2 FACTORS: 22.4 L ≈ 1 mol (gas at STP)

$0.259\ g\ Na_2CO_3 \times \dfrac{1\ mol\ Na_2CO_3}{106\ g\ Na_2CO_3} \times \dfrac{1\ mol\ CO_2}{1\ mol\ Na_2CO_3} \times \dfrac{22.4\ L\ CO_2}{1\ mol\ CO_2} = 0.0547\ L\ CO_2$

For Question 5, remember that the conversion relationship 22.4 L ≈ 1 mol applies only to a gas, and then only if that gas is under STP conditions. See Example 15.21, text p. 417.

Assignment 15D

1) GIVEN: 0.107 eq NaOH/L NaOH, 24.2 mL H_2SO_4 (0.0242 L), eq NaOH = eq H_2SO_4
   WANTED: mL NaOH, V_b,   EQUATION: $V_bN_b = V_aN_a$ where a, b refer to acid, base

   $$V_b = \frac{V_a \cdot N_a}{N_b} = \frac{(0.0242)(0.153)}{(0.107)} = 0.0346 \text{ L} = 34.6 \text{ mL}$$

   Study text pp. 419–414. Make sure you can manipulate Equation 15.15 well.

2) GIVEN: 22.73 mL H_2SO_4 (0.02273 L), 0.504 g Na_2CO_3, 106.0 g Na_2CO_3/mol
   2 eq/mol Na_2CO_3, eq Na_2CO_3 = eq H_2SO_4,   WANTED: $N_{H_2SO_4}$
   PATH: g Na_2CO_3 → mol Na_2CO_3 → eq Na_2CO_3 = eq H_2SO_4 → eq H_2SO_4/L

   $$0.504 \text{ g } Na_2CO_3 \times \frac{1 \text{ mol } Na_2CO_3}{106.0 \text{ g } Na_2CO_3} \times \frac{2 \text{ eq base}}{1 \text{ mol } Na_2CO_3} \times \frac{1 \text{ eq acid}}{1 \text{ eq base}} \times \frac{1}{0.02273 \text{ L } H_2SO_4} = 0.418 \text{ N } H_2SO_4$$

   Read text pp. 419–421, particularly Example 15.25.

3) GIVEN: 23.7 mL NaOH (0.0237 L), 25.0 mL H_3PO_4, 0.163 eq/L H_3PO_4
   WANTED: N_{NaOH}, N_1 EQUATION: $V_1N_1 = V_2N_2$

   $$N_1 = \frac{V_2N_2}{V_1} = \frac{0.0250 \text{ L}_2 \times 0.163 \text{ eq/L}_2}{0.0237 \text{ L}_1} = 0.143 \text{ N NaOH}$$

   Read the text pp. 396-398, especially Example 15.26, p. 421.

4) GIVEN: 83.1 g $C_2H_6O_2$, 62.1 g $C_2H_6O_2$/mol, 112 g H_2O, 1.86°C/m   WANTED: T_f
   EQUATION: $\Delta T_f = K_f m$

   $$\frac{83.1 \text{ g } C_2H_6O_2}{0.112 \text{ kg solvent}} \times \frac{1 \text{ mol } C_2H_6O_2}{62.1 \text{ g } C_2H_6O_2} = \frac{11.9 \text{ mol } C_2H_6O_2}{\text{kg solvent}} \text{ so m = 11.9}$$

   Substituting in Equation 15.17, $\Delta T_f = K_f m$
   $$\Delta T_f = (1.86°C/m)(11.9 \text{ m}) = 22.2 \text{ °C}$$
   $$T_f = 0°C - 22.2°C = -22.2°C$$

   Read text pp. 421–423, top, and study Example 15.27.

5) GIVEN: ΔT_f = 1.97°C, K_f = 1.86°C/m, 20.79 g (0.02079 kg) H_2O WANTED: m = mol unknown/kg H_2O
   EQUATION: $m = \frac{\Delta T_f}{K_f} = \frac{1.97°C}{1.86°C/m} = 1.06 \text{ m} = 1.06 \text{ mol unknown/kg } H_2O$

   $$\frac{3.52 \text{ g solute}/0.02079 \text{ kg } H_2O}{1.06 \text{ mol unknown/kg } H_2O} = \frac{3.52 \text{ g solute}}{0.02079 \text{ kg } H_2O} \times \frac{\text{kg } H_2O}{1.06 \text{ mol solute}} = 1.60 \times 10^2 \text{ g/mol}$$

   Study Example 15.28 on text p. 423, and the Procedure on that page.

Sage Advice and Chapter Clues

Learn well the language of solutions and use it correctly. In addition to the terms at the beginning of the chapter, know *exactly* what is meant by percentage and molarity, and, if included in your course, molality and normality. Knowing "exactly what is meant" includes the units in which these concentrations are expressed. If the units are known, understood, and used in calculation setups, you will enjoy much more success in solving solution problems. Table 15.1, text p. 414, is a big support here. Highly recommended.

When predicting solubility of one substance in another, remember that a solution is more likely to develop between substances with similar intermolecular attractions and molecular size. This is often expressed in the phrase "Like dissolves like," which suggests that similar molecules dissolve in each other. This is a common memory device, like "Oil and water don't mix." and it works most of the time.

In solving molarity and normality problems, note that both concentrations are defined in liters. If a problem gives milliliters, the volume must be converted to liters by moving the decimal 3 places left.

Do you have trouble remembering how the melting and boiling points of a solution change, compared to the pure solvent? If so, look forward in the text to Figure 21.11 on p. 584 (ICP only) and look at the label on the right-side jug of automobile anti-freeze. It says "anti-freeze and summer coolant." The anti-freeze impurity keeps the radiator contents liquid in both winter and summer. To do so, the impurity makes the melting point of the liquid decrease (no frozen engine blocks) and helps to make the boiling point increase (no geysers on hot days.)

In solving freezing point depression or boiling point elevation problems, note that the temperature term obtained is the change from the normal freezing or boiling point. This is particularly troublesome with the freezing of water. The 0.0°C freezing point of water tends to blur the difference between the new freezing point, T_f, and the term found from the equation, ΔT_f.

Do you miss the performance goals? Did you devise your own performance goals? Yes, we know it's tough to summarize. It's tough to summarize anything, not just chemistry. However, when you know how to summarize you've completed a necessary and major step in learning how to learn. We suggested you use your outline subheadings as your performance goals. If you use subheadings as performance goals, what do you use as headings? Try rewriting the section titles *using your own words*, and see where that leads.

One last thing. Go back to the floating and non-floating cola cans on text p. 35. Both colas are solutions in the cans, because the CO_2 is dissolved in the liquid. The non-floating can contains about 40 grams of sweeteners; the diet cola uses aspartame, which is 180 times sweeter than table sugar and 103 times sweeter than high fructose corn syrup. Forty grams of sugars are replaced by approximately 0.3 grams of aspartame. More water is then added to the diet cola to make up the volume to 354 mL. (The difference in mass of the two colas was 11.7 g.)

Chapter 15 Sample Test

Instructions: For each multiple choice question, select the letter of the *best* choice. Solve the remaining problems in the space provided. You may use a "clean" periodic table. Sample test questions for the optional sections on molality, normality and colligative properties follow this test.

1) A solution that can dissolve more of a substance than it currently contains is said to be _____.
   (a) supersaturated   (b) saturated   (c) unsaturated   (d) immiscible

2) When undissolved solute is present in an unsaturated solution, the rate of dissolving is _____ the rate of crystallization.
   (a) greater than   (b) equal to   (c) less than

3) The time required between adding excess solute to a solvent and reaching equilibrium with a saturated solution can be reduced by all of the following *except*:
   (a) raising the temperature   (b) stirring the mixture
   (c) illuminating the solution   (d) reducing solute particle size

4) Considering the structural formulas of CH_3OH and CH_3CH_2OH, it is logical to expect these liquids are:
   (a) miscible because their intermolecular attractions are similar.
   (b) immiscible because one molecule is highly polar and the other nonpolar.
   (c) immiscible because one molecule has hydrogen bonding and the other does not.
   (d) immiscible because their molecular masses and sizes are so different.

5) A cylinder contains liquid water and nitrogen, oxygen and carbon dioxide gases. Only carbon dioxide dissolves appreciably in the water, and there is an equilibrium between the dissolved and undissolved carbon dioxide. More nitrogen gas is forced into this cylinder, at constant temperature. The solubility of the carbon dioxide in the water _____.
   (a) increases greatly   (b) increases slightly   (c) remains unchanged   (d) decreases slightly

6) How would you make 312 grams of a 0.90% sodium chloride solution? The solvent is water.

7) How many mL of 0.415 M H_2SO_4 do you need to have 0.716 mole H_2SO_4?

8) A 62.5 mL sample of 12.0 M HNO_3 is diluted to a final concentration of 0.812 M. What is the final volume of this solution?

Questions 9–12 refer to the equation $\quad Na_2CO_3(aq) + 2\,HCl(aq) \rightarrow CO_2(g) + 2\,NaCl(aq) + H_2O(\ell)$

9) What volume of $CO_2(g)$, measured at 756 torr and 23°C, is obtained from the reaction of 24.16 mL of 0.0872 M Na_2CO_3 with excess HCl?

10) What volume of 0.123 M HCl is needed to react with 14.37 mL of 0.102 M Na_2CO_3?

11) What is the molarity of the HCl if 19.46 mL react with 0.317 grams of sodium carbonate?

12) What is the molarity of the HCl if 22.4 mL react with 11.7 mL of 0.113 M Na_2CO_3?

Check your sample test answers with those on SG pp. 213–214.

Molality and Colligative Properties Sample Test

Instructions: Solve the problems in the spaces provided. You may use a "clean" periodic table.

1) Adding 1.60 g acetic acid to 21.47 g of solvent raises the boiling point of the solution 3.14°C. What is K_b for this solvent?

2) If 5.22 grams of naphthalene, $C_{10}H_8$, are dissolved in 35.81 grams of cyclohexane, what is the molality of the resulting solution?

3) If pure cyclohexane freezes at 6.5°C and the molal freezing point constant K_f is 20.2°C/m for cyclohexane, what is the freezing point of the solution made in Question 2, above?

4) A solution is made by dissolving 4.18 g of an unknown solid in 19.89 g solvent causes the freezing point to fall by 4.31°C. If K_f for the solvent is 5.48°C/m, what is the molar mass of the unknown solid?

Check your sample test answers with those on SG pp. 214–215.

Normality Sample Test

Instructions: Solve the problems in the spaces provided. You may use a "clean" periodic table.

Questions 1–4 refer to the equation $\quad Na_2CO_3(aq) + 2\, HCl(aq) \rightarrow CO_2(g) + 2\, NaCl(aq) + H_2O(\ell)$

1) What is the equivalent mass of Na_2CO_3 in the above reaction?

2) A volume of 17.9 mL of 0.119 N Na_2CO_3 contains how many equivalents of base in the reaction above?

3) What is the normality of the HCl if 21.42 mL react with 0.288 grams of sodium carbonate in the above reaction?

4) What is the normality of the sodium carbonate if 11.3 mL react with 25.1 mL of 0.0995 N HCl in the above reaction?

Check your sample test answers with those on SG p. 215.

Chapter 16

Reactions that Occur in Water Solution: Net Ionic Equations

Did you formulate your own performance goals for Chapter 15? If not, here's a second chance to check your summarizing skills. Once again, the performance goals for this chapter are listed in the Chapter in Review section. This is the last time, though. In Chapter 17, both the performance goals and the study hints are removed, to prepare you for the next chemistry course.

Assignment 16A: Solution Conductivity and Inventories, Strong Acids and Weak Acids

Some solutes exist in water as molecules, and some solutes exist in water as ions. You need to know which solutes exist in which form to describe reactions that occur in water. Here are some ideas that lead you to this awareness:

1) A solute may be classified as a strong electrolyte, a weak electrolyte or a nonelectrolyte according to the ability of its water solution to conduct electricity.

2) If a solution **conducts** electricity, **ions** must be present as solute particles.

3) A **solution inventory** lists the ions and/or molecules that are the **solute particles** in a solution.

4) The inventory of a solution of an ionic compound is always the ions in the compound.

5) When a **strong acid** dissolves, it **dissociates into ions**. The solution inventory consists of hydrogen ions and anions.

6) There are seven strong acids. Their names and their formulas should be learned.

7) If an acid is not one of the seven strong acids, it is considered weak.

8) When a **weak acid** dissolves, it does not dissociate into ions to a large extent. The solution inventory is **mainly unionized molecules**.

Learning Procedures

Study
Sections 16.1–16.3, text pp. 435–440. Focus on your outline subheadings as you study.

Answer
Questions and Problems 1–12, text pp. 455–456.

Check your answers carefully with those on text pp. A.38–39.

Take
the skills quiz below. Check your answers with those on Study Guide (SG) p. 144.

Assignment 16A Skills Quiz

1) What property must a solute have to be classified a strong electrolyte? Weak electrolyte? Nonelectrolyte?

2) Sugar, table salt (sodium chloride), and the principal ingredient in vinegar, acetic acid, all dissolve in water. Classify each solution as a good electrical conductor, poor conductor, or nonconductor. Explain your answers.

3) Write the solution inventory for each of the following ionic compounds when it is dissolved in water:

   sodium iodide _____   barium nitrate _____

   lithium sulfate _____   ammonium phosphate _____

4) A solution of formic acid, HCO_2H, conducts electricity poorly. Explain why.

5) Write the solution inventory for each of the following acids when it is dissolved in water:

   nitric acid _____   HF _____

   hydrosulfuric acid _____   sulfuric acid _____

Assignment 16B: Net Ionic Equations

Both silver nitrate and sodium chloride are colorless, clear solutions. If equal volumes of silver nitrate and sodium chloride solutions of the same concentration are mixed, a white solid forms. (Check the picture on text p. 446.) The solid is pure silver chloride, and the ions remaining in the solution are sodium and nitrate. Can you write an equation that describes this process, identifying only the reactants and products, showing only the chemical change that has occurred?

An equation that shows precisely what happens in a solution reaction, and no more, is called a net ionic equation. To write net ionic equations you must first be able to write conventional equations for "single replacement" redox reactions, precipitation reactions, and neutralization reactions, as described on text pp. 210–216. From that starting point, you must know about the new ideas in this assignment:

1) For an ionic equation, strong acids and ionic compounds designated (aq) in a conventional equation are rewritten in their solution inventory form.

2) An ionic equation is made into a net ionic equation by removing the spectators, those species that are on *both* sides of the ionic equation.

3) Prediction of a "single replacement" redox reaction is made by referring to an activity series.

4) Prediction of an ion combination that yields a precipitate can be made from a solubility table or from solubility rules.

5) Prediction of an ion combination that yields a molecular product is made from the difference between strong and weak acids. Weak acids, including water, are formed from strong acids, but strong acids can not be formed from weak acids.

6) When ion combinations yield unstable substances, the right side of the net ionic equation has the formulas of the stable decomposition products.

Learning Procedures

Study
Sections 16.4–16.10, text pp. 440–453. Focus on your outline subheadings as you study.

Answer
Questions and Problems 13–68, text pp. 456–457. Check your answers with those on text p. A.39.

Take
the skills quiz below and on the next page. Check your answers with those on SG pp. 144–145.

Assignment 16B Skills Quiz

1) Given the activity series on the right, write the conventional, ionic, and net ionic equations for any reaction that occurs between each pair of reactants below. If no reaction occurs, write NR.

    (a) Al(s), $Ni(NO_3)_2$(aq)

Li
Ba
Na
Al
Ni
H_2
Ag

(b) Al(s), Ba(NO$_3$)$_2$(aq)

2) Write the net ionic equation for any reaction that occurs when the following pairs of reactants are combined. If you predict no reaction, write NR. You may use the solubility table on text p. 420 only if your instructor allows you to do so.

(a) solutions of nickel nitrate and sodium hydroxide

(b) solutions of sodium bromide and ammonium sulfate

(c) solutions of potassium acetate and chloric acid

(d) solutions of magnesium sulfite and hydrochloric acid

3) Give a net ionic equation for the formation of iron(II) phosphate.

Answers to Chapter 16 Skills Quiz Questions

Assignment 16A

1) To be classified as a strong electrolyte, a solute must dissociate into ions; its solution inventory will be ions. A weak electrolyte has a solution inventory of a few ions, but mainly unionized molecules. A nonelectrolyte gives a solution having only neutral molecules as the smallest particle. See text pp. 435–436.

2) Sugar solution is a nonconductor; its solution inventory is sugar molecules. Table salt releases its ions upon dissolving, and is therefore a good conductor. Acetic acid, a molecular compound, ionizes slightly in water, yielding a few ions; it is therefore a poor conductor. Read text pp. 437–440, top.

3) NaI: $Na^+(aq) + I^-(aq)$ $Ba(NO_3)_2$: $Ba^{2+}(aq) + 2\ NO_3^-(aq)$
   Li_2SO_4: $2\ Li^+ + SO_4^{2-}(aq)$ $(NH_4)_3PO_4$: $3\ NH_4^+(aq) + PO_4^{3-}(aq)$
   See Examples 16.1 and 16.2, text pp. 437–438, and Quick Check 16.2 on p. 438.

4) A formic acid solution conducts electricity poorly because formic acid dissolved in water ionizes only very slightly. The solution inventory of HCO_2H is mainly molecules, with few ions. See Example 16.3, text p. 439.

5) HNO_3: $H^+(aq) + NO_3^-(aq)$ HF: $HF(aq)$
   H_2S: $H_2S(aq)$ H_2SO_4: $2\ H^+(aq) + SO_4^{2-}(aq)$
   Read text pp. 438–440, especially Example 16.3, p. 439; don't overlook Quick Check 16.3, top of text p. 440.

Assignment 16B

1) For Problem 1, study text pp. 441–444. For (a), see Example 16.4, text p. 416; for (b), see Example 16.6, text p. 443.
   (a) $2\ Al(s) + 3\ Ni(NO_3)_2(aq) \rightarrow 2\ Al(NO_3)_3(aq) + 3\ Ni(s)$
   $2\ Al(s) + 3\ Ni^{2+}(aq) + 6\ NO_3^-(aq) \rightarrow 2\ Al^{3+}(aq) + 6\ NO_3^-(aq) + 3\ Ni(s)$
   $2\ Al(s) + 3\ Ni^{2+}(aq) \rightarrow 2\ Al^{3+}(aq) + 3\ Ni(s)$
   (b) $2\ Al(s) + 3\ Ba(NO_3)_2(aq) \rightarrow$ NR   Al is below Ba in the activity series, so it will *not* replace Ba^{2+} ions in solution. See Table 16.1, text p. 443.

2) For Problem 2, study text pp. 444–453. Specific examples and page references from the text are given at the end of each question.
   (a) $Ni^{2+}(aq) + 2\ OH^-(aq) \rightarrow Ni(OH)_2(s)$ 16.9–16.10, pp. 445–446
   (b) $Na^+(aq) + Br^-(aq) + NH_4^+(aq) + SO_4^{2-}(aq) \rightarrow$ NR   16.15, p. 449
   (c) $C_2H_3O_2^-(aq) + H^+(aq) \rightarrow HC_2H_3O_2(aq)$ 16.18, p. 450
   (d) $SO_3^{2-}(aq) + 2\ H^+(aq) \rightarrow SO_2(g) + H_2O(\ell)$ pp. 451–452

3) $3\ Fe^{2+}(aq) + 2\ PO_4^{3-}(aq) \rightarrow Fe_3(PO_4)_2(s)$ Example 16.14, pp. 448–449.

Sage Advice and Chapter Clues

The ability to write correct net ionic equations is probably the most important predictor of success in general chemistry, especially the second term. Consequently, the advice in this chapter is long, to make certain you understand this topic.

Start with Summary Table 16.4, text p. 453. You should be able to "sort" these reactions into types (column two of the table) and from each type, predict the products (column four of the table.) If you know solution inventories, you can easily obtain column six of the table from column four.

To write correct solution inventories, you must be able to predict electrical conductivities of aqueous solutions. The table below summarizes the relationships between electrolytes, solution inventories, electrical conductivity and solutes.

Electrolyte	Solution Inventory	Conductivity	Solute Types
Strong	Nearly 100% ions	Good	All soluble salts Strong acids
Weak	Mostly molecules, plus a few ions	Poor	Weak acids
Nonelectrolyte	Essentially 100% molecules	Nonconductor	Unionized compounds

---

Net ionic equations is another of those topics where, if you develop a systematic approach, you will have little difficulty. Here are some very strong suggestions about writing net ionic equations.

1) Follow closely the three step procedure in the Summary on page 441 in the text. Students who don't attempt to shortcut the procedure in the learning process are far more successful than those who do. Put another way, *ALL* students who have trouble have one thing in common: they shortcut the procedure.

2) For each species in the conventional equation (Step 1, text p. 415), write the state designation, (g), (ℓ), (s) or (aq). This is very important.

3) In writing the ionic equation (Step 2), repeat all species designated (g), (ℓ), (s) or (aq) in the conventional equation in exactly the same form. Do not change them.

4) Be sure you know the difference between strong and weak acids, and can tell whether a given acid is strong or weak. Study pp. 438–440 in the text. All the acids you'll meet this term are soluble, so don't sweat acid solubilities. All the acids dissolve, the important question is "Do they ionize?"

5) If a species designated (aq) in the conventional equation is an ionic compound or a strong acid, rewrite that species with separate ions in the ionic equation (Step 2). If a species designated (aq) in the conventional equation is a weak acid, repeat it in the ionic equation in exactly the same form, not separated into ions. These two instructions are the direct consequences of the first two entries in the solute types column of the table above.

6) When you eliminate spectators from the ionic equation to make the net ionic equation (Step 3), do just that. Eliminate only those things that appear on both sides of the equation; don't change anything else.

7) Be sure the net ionic equation is balanced both in atoms and electrical charge.

8) Reduce the net ionic equation to lowest terms by dividing all coefficients by a common factor, if any.

9) When a strong acid like HCl ionizes, it gives the hydrogen ion, H^+. When a strong acid like H_2SO_4 ionizes, it also produces the hydrogen ion, H^+. It does not produce an H_2^{2+} ion, but rather $2\,H^+(aq)$.

Chapter 16 Sample Test

Instructions: You may use a "clean" periodic table; you may use a solubility table only if your instructor allows. For Question 1, select the letter of the best choice; for Questions 2-8, put the answers in the spaces provided.

1) A soluble salt is a _____ electrolyte.
   (a) weak   (b) non   (c) strong

2) Write solution inventories of the following substances:

   calcium chloride _____   hydroiodic acid _____

   acetic acid _____   aluminum nitrate _____

Write the net ionic equation for each reaction below and on the next page. Write NR if you predict no reaction.

3) Solutions of sodium sulfate and potassium carbonate are mixed.

4) Hydrochloric acid is added to ammonium carbonate solution.

5) Sodium formate, $NaCHO_2$, solution is added to sulfuric acid.

6) Lead(II) nitrate and nickel sulfate solutions are mixed.

7) Solid calcium hydroxide is treated with nitric acid.

8) Hydrochloric acid is poured on nickel metal (hydrogen gas is below nickel metal in the activity series).

Check your sample test answers with those on SG p. 215.

CHAPTER

17

Acid-Base (Proton Transfer) Reactions

Assignment 17A: Acid-Base Theories

Acid-base reactions are everywhere around you. They include small scale reactions such as controlling the acidity of your blood and body cells and adding lemon to a cup of tea. They also include large scale industrial processes that consume billions of kilograms of chemicals per year. "Acid rain" is a sad example of the importance of acid-base reactions in our lives. Indeed, acid-base reactions are among the most important classes of reactions; they surely deserve detailed study.

The main ideas that you should learn from this assignment are:

1) The **hydrogen ion-hydroxide ion** concept of acids and bases is based on chemical properties and is known as the **Arrhenius theory** of acids and bases.

2) According to the **Brönsted-Lowry** theory, an acid-base reaction involves a transfer of a proton from one substance, the acid, to another, the base. An **acid** is a **proton donor**; a **base** is a **proton acceptor**. When writing an equation, there must be a Brönsted-Lowry base whenever there is a Brönsted-Lowry acid.

3) Acid-base reactions are reversible; they reach a state of equilibrium according to the **general equation** $HA + B \rightleftarrows A^- + HB^+$.

4) According to the **Lewis** theory of acids and bases, an **acid** is an **electron pair acceptor** and a **base** an **electron pair donor**.

5) Two substances whose formulas differ only by a proton (a hydrogen ion) are a **conjugate acid-base pair**. If the **acid** has the general form **HA**, its **conjugate base** has the general form **A⁻**. If the **base** has the general form **B**, its **conjugate acid** has the general form **HB⁺**.

Learning Procedures

Study
Sections 17.1–7.5, text pp. 460–467. Focus on your outline subheadings as you study.

Answer
Questions and Problems 1–10, text p. 482. Check your answers with those on text p. A.39.

Take
the skills quiz below and on the next page. Check your answers with those on Study Guide p. 153.

Assignment 17A Skills Quiz

1) The electrolyte in a car battery is a sulfuric acid solution. What would happen if:
   (a) this electrolyte spilled on the car's active metal frame?

   (b) an unscrupulous mechanic dropped an Alka-Seltzer™ tablet (sodium hydrogen carbonate) into a car battery?

   (c) battery electrolyte were mixed with a litmus indicator solution?

2) What ion in a car battery's electrolyte is responsible for all the chemical reactions in Question 1, above?

3) Define a Brönsted-Lowry acid.

4) From the following, state which are potential Lewis acids and which are potential Lewis bases: Ca^{2+}, F^-, Na^+, CHO_2^-, Fe^{3+}, NH_3.

5) Identify Brönsted-Lowry conjugate acid-base pairs in the equation:
$$HC_2H_3O_2(aq) + HPO_4^{2-}(aq) \rightleftarrows C_2H_3O_2^-(aq) + H_2PO_4^-(aq)$$
Identify the Brönsted-Lowry acid in each pair, and the base in each pair.

Assignment 17B: Relative Acid Strengths, Predicting Acid-Base Reactions

This assignment examines Brönsted-Lowry acid-base reactions in some detail. The following ideas are introduced in this short lesson:

1) **Strong** acids transfer protons **readily**; weak acids do not. Strong bases accept protons readily; weak bases do not.

2) The relative strengths of Brönsted-Lowry acids and bases may be decided from their positions in a table.

3) Proton transfer reactions involve two conjugate acid-base pairs. When a proton transfer reaction reaches **equilibrium**, the proton transfer yields the **weaker conjugate acid** and the **weaker conjugate base**.

Learning Procedures
Study
Sections 17.6–17.7, text pp. 467–471. Focus on your outline subheadings as you study.

Answer
Questions and Problems 19–34, text p. 483. Check your answers with those on text p. A.39.

Take
the skills quiz below. Check your answers with those on SG p. 153.

Assignment 17B Skills Quiz

Instructions: You may use Table 17.1, text p. 468, during this skills quiz.

1) An acid solution has a solution inventory that is mainly molecular. Is this acid a strong acid or a weak acid?

2) If HNO_3 is a stronger acid than HF, which of NO_3^- and F^- is the stronger base?

3) Complete the equation below for a proton transfer reaction. Using Table 17.1, predict which direction, forward or reverse, is favored as the reaction reaches equilibrium.

$$HC_2H_3O_2(aq) + HS^-(aq) \rightleftarrows$$

Assignment 17C: The Water Equilibrium and pH Conversions

For this assignment, you may need to review exponents and logarithms. Use Appendix I parts B (text pp. A.6,7) and C (text pp. A.8,9) for this review.

Look for the main ideas in this assignment:

1) Water itself is *both* a weak acid *and* a weak base.

2) Because concentrations of $H^+(aq)$ or $OH^-(aq)$ are usually small, but can vary over wide ranges, they are usually expressed in exponential notation and as logarithms.

3) **pH = -log[H$^+$]; pOH = -log[OH$^-$]**. From these equations, you can obtain $[H^+] = 10^{-pH}$, and $[OH^-] = 10^{-pOH}$. In water solutions at 25°C, **pH + pOH = 14.00**.

4) A **neutral** water solution has **pH = 7.00** at 25°C. An **acidic** water solution has **pH less than 7.00** at 25°C; a **basic** water solution has **pH greater than 7.00**, also measured at 25°C.

Section 17.10, using noninteger pH conversions, is an optional section. The new skills needed to complete this optional section are not chemical, but mathematical. The ideas concerning pH, pOH, $[H^+]$ and $[OH^-]$ are the same as in Section 17.9; only the numbers have been changed.

Learning Procedures

Study
Sections 17.8–17.9, text pp. 471–477. If required, also study Section 17.10, text pp. 477-480. Focus on your outline subheadings as you study.

Answer
Questions and Problems 35–52, text pp. 483–484. If required, also do the optional 53–60 on text p. 484. Check your answers with those on text pp. A.39.

Take
the skills quiz below and on the next page. Check your answers with those on SG pp. 153–154.

Assignment 17C Skills Quiz

1) Write an equation describing the ionization of water.

2) Match the letters of the appropriate choices:

   acidic solution_____   neutral solution_____   basic solution_____

   (a) pH = 7.00   (b) $[H^+]$ less than 10^{-7} (c) pH less than 7.00

   (d) $[H^+]$ greater than 10^{-7} (e) pH greater than 7.00   (f) $[H^+] = 10^{-7}$

3) A solution has a pH = 3.00. Calculate the [H⁺], [OH⁻], and pOH of this solution.

4) Solution A has pH = 1.00; solution B has pH = 9.00; solution C has pH = 4.00; solution D has pH = 11.00. Arrange these solutions in order of *increasing* basicity (most basic first).

Questions 5 and 6 are from the optional Section 17.10.

5) The hydrogen ion concentration of a solution is 6.2×10^{-5}. Calculate the pH, pOH and [OH⁻] of this solution.

6) The pOH of a solution is 3.47. Calculate the pH, [H⁺] and [OH⁻] of this solution.

Answers to Chapter 17 Skills Quiz Questions

Assignment 17A

1) (a) Hydrogen gas would be evolved, and the metal frame would dissolve. See text p. 460, and review pp. 441–444.
   (b) Carbon dioxide gas would be evolved—and the car might not start!
   (c) The litmus indicator would turn red.

   For parts (b), (c) check p. 460 in the text.

2) The hydrogen ion, $H^+(aq)$, is responsible for all the reactions in Question 1. Read text pp. 460–461.

3) A Brönsted-Lowry acid is a proton transfer agent. See text pp. 461–464. The most important word here is *transfer*; you can't have a Brönsted-Lowry acid without a Brönsted-Lowry base, something to receive the transferred proton.

4) The potential Lewis acids are Ca^{2+}, Na^+ and Fe^{3+}; the potential Lewis bases are F^-, CHO_2^- and NH_3. Check text pp. 464–465, and don't forget about the lone pair of electrons in the Lewis diagram for NH_3.

5) The conjugate acid-base pairs are $HC_2H_3O_2$—$C_2H_3O_2^-$ and $H_2PO_4^-$—HPO_4^{2-}. The conjugate acids are $HC_2H_3O_2$ and $H_2PO_4^-$; the conjugate bases are $C_2H_3O_2^-$ and HPO_4^{2-}. Study text pp. 465–467 and Examples 17.1–17.3.

Assignment 17B

1) If the solution inventory is mainly molecular, the acid is a weak acid, because it does not ionize much in water. Read text p. 468 and review pp. 438–440.

2) The fluoride ion, F^-, is a stronger base than NO_3^-. The weaker the acid, the stronger the conjugate base. Read text pp. 467–469, and study Examples 17.4 and 17.5. Note the trends between position in Table 17.1 and acid, base strength.

3) The completed equation is:
$$HC_2H_3O_2(aq) + HS^-(aq) \rightarrow C_2H_3O_2^-(aq) + H_2S(aq).$$
From Table 17.1, H_2S is a weaker acid than $HC_2H_3O_2$ and $C_2H_3O_2^-$ is a weaker base than HS^-. The right side of this equation has both the weaker acid and the weaker base, so the favored position is to the right. Figure 17.2, text p. 469, depicts visually the previous two sentences. Study text pp. 469–471 and Examples 17.6-17.7.

Assignment 17C

1) The equation for water ionization is $HOH(\ell) \rightleftarrows H^+(aq) + OH^-(aq)$. Check p. 472 in the textbook for Equation 17.9.

2) Acidic solution: c, d; neutral solution: a, f; basic solution: b, e. Read the text pp. 471–473.

3) If pH = 3.00, pOH = 14.00 - 3.00 = 11.00 $[H^+] = 10^{-pH} = 10^{-3}$ $[OH^-] = 10^{-pOH} = 10^{-11}$
   Read text pp. 443-447 and see Examples 17.9–17.11.

4) The solution with the largest pH value is the most basic, so in order of decreasing basicity, D > B > C > A. See Example 17.12, text p. 476.

Questions 5 and 6 are from the optional Section 17.10. Read pp. 477–480.

5) pH = $-\log(6.2 \times 10^{-5})$ = $-(-4.21)$ = 4.21
   pOH = 14.00 - pH = 14.00 - 4.21 = 9.79 $[OH^-] = 10^{-(pOH)} = 10^{-9.79} = 1.6 \times 10^{-10}$
   Study Examples 17.13–17.17, text pp. 478–480. Be sure you understand the procedure needed to enter logarithms and inverse logarithms on your calculator. This advice applies to Question 6 also for the math is basically the same.

6) pH = 14.00 - pOH = 14.00 - 3.47 = 10.53
   $[H^+] = 10^{-pH} = 10^{-10.53} = 3.0 \times 10^{-11}$
   $[OH^-] = K_w / [H^+] = 1.0 \times 10^{-14} / 3.0 \times 10^{-11} = 3.3 \times 10^{-4}$. See Question 5, above.

Sage Advice and Chapter Clues

Don't be worried by the existence of three different acid-base systems. All three systems generally agree on which species in a reaction is the acid and which is the base. It's not a question of one system being right and the others wrong, but rather which system is most convenient for the reaction being studied. Both good workers and good students know how to pick the best tool for the job at hand.

Brönsted-Lowry acid-base theory is most convenient for proton transfer reactions. Correctly written equations show two conjugate acid-base pairs, and the position of equilibrium favors the formation of the weaker acid and the weaker base. From Table 17.1, text p. 468, you can predict relative acid and base strength, so you can predict if a proton transfer reaction will occur. Because the weaker acid and base in a reaction are less highly ionized than the stronger acid and base, this is another example of an ion-combination reaction forming a "more" molecular product (Section 16.7, text pp. 449–451. Be sure you can do the pH loop given in Figure 17.3, text p. 474.

As long as we're talking about pH calculations, some words concerning calculators are in order. If you understand what you are doing mathematically when you push buttons on a calculator, that little number box can be your best friend. If you don't understand what's going on when you push the "log" or the "inv" button, the calculator can be your worst enemy. Why? Most errors committed when using calculators are typing errors; data are entered incorrectly on the keyboard. The calculator then quickly gives an incorrect, but impressive, answer. If you have a feel for the size of the answer, you can tell many times when you've punched in the wrong data. If you don't, the calculator quickly leads you down the primrose path to numeric oblivion.

A memory aid for Lewis acids and bases is: Lewis *a*cid=electron *a*cceptor. Most cations are Lewis acids, and anions tend to be Lewis bases. All Lewis bases have unshared electron pairs. All Brönsted-Lowry bases are also Lewis bases, but H$^+$(aq) is only one of many Lewis acids.

Chapter 17 Sample Test

Instructions: Select the letter of the *best* choice for Questions 1–11. For Questions 12–16, solve the problem in the space provided. You may use Table 17.1, text p. 468, during this sample test.

1) Which of these properties is *not* related to the H$^+$(aq) ion?
   (a) turns litmus indicator red
   (b) slippery feeling on the skin
   (c) sour taste
   (d) reacts with and neutralizes a base

2) According to the Brönsted-Lowry acid-base theory, an acid is a(n)
   (a) electron donor   (b) electron acceptor   (c) proton donor   (d) proton acceptor

3) According to the Lewis acid-base theory, a base is a(n)
   (a) electron donor   (b) electron acceptor   (c) proton donor   (d) proton acceptor

4) Identify the Lewis acid among the following:
   (a) NH_3 (b) Cl^- (c) O_2 (d) K^+

5) For the reaction $HX^- + HY^- \rightleftarrows H_2X + Y^{2-}$, pick the conjugate acid-base pair among the following:
   (a) HX^-, HY^- (b) HX, Y^{2-} (c) HY^-, H_2X (d) Y^{2-}, HY^- (e) HX^-, Y^{2-}

6) The stronger of two Brönsted-Lowry bases tends to:
   (a) lose protons more easily
   (b) lose protons less easily
   (c) gain protons more easily
   (d) gain protons more easily

7) Using Table 17.1, text p. 468, identify the strongest acid among the following:
   (a) HF   (b) SO_3^{2-} (c) NH_4^+ (d) $H_2PO_4^-$

8) At 25°C, the numeric value of K_w is:
   (a) 1.0×10^{-7} (b) 1.0×10^{-14} (c) 1.0×10^{14} (d) 14.00

9) A solution with a pH of 7.6 would be classified as:
   (a) strongly acidic   (b) about neutral   (c) strongly basic

10) In a solution in which $[OH^-] = 10^{-5}$, pH and pOH are, respectively,
    (a) 10^{-14} and 10^{-9} (b) 5 and 9   (c) 5 and 14   (d) 9 and 5

11) Which of the following solutions is most basic?
    (a) $[OH^-] = 10^{-5}$ (b) pH = 7   (c) pOH = 9   (d) $[H^+] = 10^{-2}$

12 and 13: For each pair of reactants below, complete the equation for a single proton transfer reaction; state if the reaction would be favored in the forward or the reverse direction. You may use Table 17.1, p. 468.

Favored Direction

12) $HCN(aq) + SO_3^{2-}(aq) \rightleftarrows$

13) $HSO_3^-(aq) + HPO_4^{2-}(aq) \rightleftarrows$

Questions 14–16 from optional Section 17.10 refer to a solution of pH 2.87.

14) The pOH of this solution is _____.

15) The $[H^+]$ of this solution is _____.

16) The $[OH^-]$ of this solution is _____.

Check your sample test answers with those on SG p. 216.

CHAPTER 18

Oxidation-Reduction (Electron Transfer) Reactions

Assignment 18A: Oxidation, Reduction, Redox. What Do They Mean?

If you have studied Chapter 17, you know that a proton is transferred from an acid to a base in an acid-base reaction (text pp. 461–464.) In a notably similar way, an electron is transferred from a reducing agent to an oxidizing agent in an oxidation-reduction reaction. It is easier to understand both systems if you see their similarity.

Look for the following big ideas in this assignment:

1) **Oxidation** is a **loss of electrons**, or an increase in oxidation number. **Reduction** is a **gain of electrons**, or a reduction in oxidation number.

2) Oxidation-reduction (redox) reactions can be divided into an oxidation half-reaction equation and a reduction half-reaction equation. Addition of these equations gives a balanced equation for a redox reaction.

3) In a redox reaction, the **reducing agent** (reducer) **loses electrons** and the **oxidizing agent** (oxidizer) **gains electrons**.

Learning Procedures

Study
Sections 18.1–8.4, text pp. 487–496. Focus on your outline subheadings as you study.

Answer
Questions and Problems 1–24, text pp. 505–506. Check your answers with those on text p. A.40.

Take
the skills quiz below and on the next page. Check your answers with those on Study Guide (SG) p. 162.

Assignment 18A Skills Quiz

1) Under what conditions is a rechargeable nickel-cadmium cell a voltaic cell? Under what conditions is it a electrolytic cell?

2) In terms of electrons gained or lost, define both oxidation and reduction.

3) Classify each of the following equations as an oxidation half-reaction equation or a reduction half-reaction equation:

(a) $ClO_2(aq) + e^- \rightarrow ClO_2^-(aq)$ _____

(b) $2\,I^-(aq) \rightarrow I_2(aq) + 2\,e^-$ _____

4) Combine the half-reaction equations in Question 3, above, to give a balanced redox reaction.

5) Give the oxidation state of the element whose symbol is **boldface** in each formula:

Ca^{2+}_____   As**O**$_2^-$_____   Al$_2$**S**$_3$_____   **N**H$_4$Br_____

6) In each reaction, identify the element oxidized or reduced and note the change in oxidation number. For example, in $Pb^{2+}(aq) + 2\,e^- \rightarrow Pb(s)$, lead is reduced from 2+ to 0.

(a) $K(s) \rightarrow K^+(aq) + e^-$ _____

(b) $2\,H^+(aq) + ClO_4^-(aq) \rightarrow ClO_3^-(aq) + H_2O(\ell)$ _____

7) In the reaction $Cu^{2+}(aq) + Mg(s) \rightarrow Cu(s) + Mg^{2+}(aq)$, identify:

(a) the substance oxidized _____   (b) the oxidizing agent _____

(c) the substance reduced _____   (d) the reducing agent _____

8) Place the letter of the choice below that *best* describes each of the following:

strong attraction for electrons _____   weak attraction for electrons _____

gives up electrons weakly _____   gives up electrons freely _____

(a) strong oxidizing agent
(b) weak oxidizing agent
(c) strong reducing agent
(d) weak reducing agent

Assignment 18B: Predicting Redox Reactions From Strengths of Oxidizing and Reducing Agents

You can tell an oxidizer from a reducer, but so far you cannot tell if a given oxidizer will accept electrons from a given reducer. This assignment will teach you how to predict if a redox reaction will occur when a given oxidizer and reducer are mixed. The new ideas in this assignment precisely parallel the acid-base concepts given on page @@@ of this study guide.

1) Strong **oxidizing agents attract electrons** strongly; weak oxidizers do not. Strong **reducers release electrons** readily; weak reducing agents do not release electrons readily.

2) The relative strengths of oxidizing and reducing agents may be determined from their relative positions in a table.

3) When a redox reaction reaches **equilibrium**, the **weaker oxidizer** and the **weaker reducer** are favored.

Learning Procedures

Study
Sections 18.5–18.7, text pp. 496–500. Focus on your outline subheadings as you study.

Answer
Questions and Problems 25–36, text p. 506. Check your answers with those on text p. A.40.

Take
the skills quiz below and on the next page. Check your answers with those on SG pp. 162–163.

Assignment 18B Skills Quiz

Instructions: You may refer to Table 18.2, text p. 496 during this quiz.

1) Which is the stronger oxidizing agent, $Br_2(\ell)$ or $Zn^{2+}(aq)$?

2) Which is the stronger reducer, $I^-(aq)$ or $Fe(s)$?

3) For the reactants $Br_2(\ell)$ and $Na(s)$, select from Table 18.2 the needed half-reaction equations, write them so they may be added to produce the equation for the redox reaction, and complete the addition. Predict the direction in which the reaction is favored.

4) For the reactants NO_3^-(aq) and Cl^-(aq), select from Table 18.2 the needed half-reaction equations, write them so they may be added to produce the equation for the redox reaction, and complete the addition. Predict the direction in which the reaction is favored.

5) Put the letter of the choice below that *best* describes each of the following:

   reducing agent_____ oxidizing agent_____ acid_____ base_____
   (a) proton donor (b) electron acceptor (c) electron donor (d) proton acceptor

---

Assignment 18C: Writing Redox Equations

You have learned how to combine half-reaction equations to get a redox equation. But you haven't learned how to develop a half-reaction equation, knowing only the beginning and final forms of the element oxidized or reduced.

It is quite easy to write a half-reaction equation when the element oxidized or reduced is not combined with another element, either as a compound or ion. When the element oxidized or reduced is so combined, however, a certain procedure must be followed. Learning that procedure is the single new idea in this assignment.

Learning Procedures
Study
Section 18.8, text pp. 500–503. Focus on your outline subheadings as you study.

Answer
Questions and Problems 37–46, text pp. 506–507. Check your answers with those on text p. A.40.

Take
the skills quiz on the next page. Check your answers with those on SG p. 163.

Assignment 18C Skills Quiz

1) Write separate oxidation and reduction half-reaction equations for the redox reaction
$$Fe^{2+}(aq) + NO_3^-(aq) \rightarrow Fe^{3+}(aq) + NO_2(g)$$
which occurs in acidic solution. Add the half-reaction equations to produce a balanced redox equation.

2) Write separate oxidation and reduction half-reaction equations for the redox reaction
$$C_2O_4^{2-}(aq) + MnO_4^-(aq) \rightarrow CO_2(g) + Mn^{2+}(aq)$$
which occurs in acidic solution. Add the half-reaction equations to produce a balanced redox equation.

Answers to Chapter 18 Skills Quiz Questions

Assignment 18A

1) When the "ni-cad" battery is powering the beeper, calculator, computer, whatever, it is acting as a voltaic cell because it is a source of electrical power. When the battery is in its charger and being recharged, it is acting as an electrolytic cell; the electric power from the wall outlet is the outside source that "forces" the ni-cad to recharge. See text pp. 487–488.

2) Oxidation is a loss of electrons; reduction is a gain of electrons. Read page 488 in the text.

3) (a) The equation depicts reduction; the reactant gains an electron.
   (b) The equation depicts oxidation; the reacting ions lose electrons.
   Study text pp. 488–491 particularly Example 18.2, text p. 490.

4)
$$2\,[ClO_2(aq) + e^- \rightarrow ClO_2^-(aq)]$$
$$\underline{2\,I^-(aq) \rightarrow I_2(aq) + 2\,e^-}$$
$$2\,ClO_2(aq) + 2\,I^-(aq) \rightarrow 2\,ClO_2^-(aq) + I_2(aq)$$

See Examples 18.2–18.3 text pp. 490–491.

5) Ca^{2+}, 2+; AsO_2^-, 3+; Al_2S_3, 2−; NH_4Br, 3−
   Read text pp. 491–493 and study Examples 18.4–18.5. If you had trouble with the N in NH_4Br, read carefully rule 2 on text p. 494.

6) (a) Potassium is oxidized from 0 to 1+.   Read text pp. 494–495.
   (b) Chlorine is reduced from 7+ to 5+.   See Example 18.6, p. 494.

7) (a) the substance oxidized, Mg(s)   (b) the oxidizing agent, Cu^{2+}(aq)
   (c) the substance reduced, Cu^{2+}(aq)   (d) the reducing agent, Mg(s)
   Study text pp. 495–496, particularly Example 18.7.

8) strong attraction for electrons, a   weak attraction for electrons, b
   gives up electrons weakly, d   gives up electrons freely, c
   Read text pp. 496–497.

Assignment 18B

1) The stronger oxidizer is $Br_2(\ell)$. 2) The stronger reducer is Fe(s). For Questions 1 and 2, read text pp. 497–499, top. Study the *patterns* of relative oxidizer and reducer strengths in Table 18.2, text p. 496.

3)
$$2\,Na(s) \rightleftarrows 2\,Na^+(aq) + 2\,e^-$$
$$\underline{Br_2(\ell) + 2\,e^- \rightleftarrows 2\,Br^-(aq)}$$
$$2\,Na(s) + Br_2(\ell) \rightleftarrows 2\,Na^+(aq) + 2\,Br^-(aq)$$

The forward reaction is favored.

Study Figure 18.2, p. 498 and Table 18.2, p. 496. Check Examples 18.8–18.12, text pp. 497–500.

Oxidation-Reduction (Electron Transfer) Reactions   163

4)
$$2\ NO_3^-(aq) + 8\ H^+(aq) + 6\ e^- \rightleftarrows 2\ NO(g) + 4\ H_2O(\ell)$$
$$6\ Cl^-(aq) \rightleftarrows 3\ Cl_2(g) + 6\ e^-$$
$$\overline{2\ NO_3^-(aq) + 8\ H^+(aq) + 6\ Cl^-(aq) \rightleftarrows 2\ NO(g) + 4\ H_2O(\ell) + 3\ Cl_2(g)}$$

The reverse reaction is favored. Again study Figure 18.2 and Table 18.2; pay special attention to Examples 18.9 and 18.10, text pp. 498–499. Read the caption to Figure 18.2 *very* carefully; make sure you understand what it says.

5)   reducing agent c   oxidizing agent b   acid a   base d

Assignment 18C

1)
$$Fe^{2+}(aq) \rightarrow Fe^{3+}(aq) + e^-$$
$$e^- + NO_3^-(aq) + 2\ H^+(aq) \rightarrow NO_2(g) + H_2O(\ell)$$
$$\overline{Fe^{2+}(aq) + NO_3^-(aq) + 2\ H^+(aq) \rightarrow Fe^{3+}(aq) + NO_2(g) + H_2O(\ell)}$$

See text pp. 500–503, and Example 18.13, text pp. 501–502; reread the Summary on p. 501.

2)
$$5\ C_2O_4^{2-}(aq) \rightarrow 10\ CO_2(g) + 10\ e^-$$
$$2\ MnO_4^-(aq) + 16H^+(aq) + 10\ e^- \rightarrow 2\ Mn^{2+}(aq) + 8\ H_2O(\ell)$$
$$\overline{5\ C_2O_4^{2-}(aq) + 2\ MnO_4^-(aq) + 16\ H^+(aq) \rightarrow 10\ CO_2(g) + 2\ Mn^{2+}(aq) + 8\ H_2O(\ell)}$$

Text pp. 500–503, Example 18.14, p. 503; study the Summary on p. 501 one more time.

Sage Advice and Chapter Clues

The vocabulary of redox reactions and equations is sometimes confusing because the terms are so closely related. The following summary should help:

New Redox Term	Change in Electrons	Change in Oxidation Number
oxidation	loss of electrons	increase
reduction	gain of electrons	decrease
oxidizing agent *or* oxidizer	accepts electrons	decrease*
reducing agent *or* reducer	donates electrons	increase*

*monatomic species

The key fact in combining half-reaction equations to get a redox reaction equation is that the total number of electrons lost by one species must be gained by the other. The half-reaction equations must be adjusted so that, when added, the electrons cancel out. If you must write a half-reaction equation in which the element oxidized or reduced is combined with another element on either side of the equation, follow exactly the procedure on p. 501 of the text. The horror stories you might have heard about balancing redox equations come from people who don't follow a set procedure as on p. 501.

Chapter 18 Sample Test

Instructions: Answer each question in the space provided.

Use the following half-reactions for Questions 1 and 2:

$Ca(s) \rightarrow Ca^{2+}(aq) + 2\ e^-$ _____

$Al^{3+}(aq) + 3\ e^- \rightarrow Al(s)$ _____

1) In the spaces to the right of each half-reaction equation above, identify the reduction half-reaction equation, and identify the oxidation half-reaction equation.

2) In the space below, rewrite the half-reaction equations in such form that they may be added to produce a balanced redox equation. Write the balanced redox equation below the half-reaction equations.

Use the following redox equation when answering Questions 3–5:
$$2\ CrO_4^{2-}(aq) + 16\ H^+(aq) + 3\ Cu(s) \rightarrow 2\ Cr^{3+}(aq) + 3\ Cu^{2+}(aq) + 8\ H_2O(\ell)$$

3) The element that loses electrons is _____. Its oxidation number changes from _____ to _____.

4) The element reduced is _____. Its oxidation number changes from _____ to _____.

5) The oxidizing agent is _____; the reducing agent is _____.

For Question 6, circle the letter of the *best* choice.

6) If element Q is very likely to form Q^- ions in a redox reaction, element Q is a _____ agent.
   (a) strong reducing   (b) strong oxidizing   (c) weak reducing   (d) weak oxidizing

Use the table of relative strengths of oxidizers and reducers to the right for Questions 7 and 8. In this table, the strongest oxidizer is W.

$$W \rightarrow W^-$$
$$X^+ \rightarrow X$$
$$Y^+ \rightarrow Y$$
$$Z \rightarrow Z^-$$

7) Identify the stronger oxidizer between Y^+ and X^+; identify the stronger reducer between W^- and Y.

8) Using the table at the top of this page, predict which of the following redox reactions will proceed in the forward direction. Circle the letter(s) corresponding to those equations.
   (a) $Z + X \rightleftarrows Z^- + X^+$
   (b) $X^+ + Y \rightleftarrows X + Y^+$
   (c) $Z^- + W \rightleftarrows Z + W^-$
   (d) $Y + Z \rightleftarrows Y^+ + Z^-$
   (e) $X^+ + W^- \rightleftarrows X + W$

9) For the redox reaction $ClO_3^-(aq) + I^-(aq) \rightarrow I_2(aq) + Cl^-(aq)$,
   (a) write the oxidation half-reaction equation

   (b) write the reduction half-reaction equation

   (c) add the half-reaction equations to produce a balanced redox equation.

Check your sample test answers with those on SG p. 216.

CHAPTER 19
Chemical Equilibrium

Assignment 19A: Chemical Equilibrium. What Is It?

Virtually everything you encounter, from your own bodily processes and senses, to the color and smell of a flower, to the formation of mountains is the result of one or more equilibrium reactions. Equilibrium exists, you will recall from text pp. 371 and 401, when reversible reactions occur at equal rates. In this chapter, we look at chemical equilibria more closely.

Look for the following big ideas.

1) Chemical **equilibria** are dynamic. The opposing **rates are equal**, but they are not zero, even though you see no visible changes. Equilibria exist only in "closed" systems. In a closed system, no matter is lost.

2) All **chemical reactions** start with **molecular collisions**, but not all molecular collisions give a chemical reaction.

3) An **energy-reaction graph** shows potential energies of reactants, activated complex and products in a reaction, and the activation energy as the reaction proceeds in either direction.

4) Reaction rates are higher at higher temperatures because a larger fraction of the sample has enough kinetic energy to participate in reaction-producing collisions. The energy of collision must be enough to overcome the mutual repulsion of each reactant's valence electrons.

5) A **catalyst** increases reaction rate by providing an alternative reaction path with a **lower activation energy**.

6) Collision rates are higher at higher concentrations, so reaction rates are higher at higher reactant concentrations.

Learning Procedures
Study
Sections 19.1–19.5, text pp. 509–515. Focus on your outline subheadings as you study.

Answer
Questions and Problems 1–24, text pp. 531–532. Check your answers with those on text p. A.41.

Take
the skills quiz below and on the next page. Check your answers with those on SG p. 172.

Assignment 19A Skills Quiz

Instructions: Select the letter of the *best* choice.

1) Which is the *correct* statement about equilibrium systems?
   (a) Equilibrium systems describe only physical changes.
   (b) Equilibrium systems describe only chemical changes.
   (c) Forward and reverse reaction rates are the same at equilibrium.
   (d) The reversible changes in an equilibrium may sometimes stop, but only if both the forward and reverse reaction stop together.

2) Which is the *incorrect* statement about collision theory?
   (a) Not all collisions give a chemical reaction.
   (b) The number of effective collisions is the same as the total number of collisions.
   (c) A collision with too little kinetic energy does not give any product.
   (d) Reaction rate depends on the frequency of effective collisions.

3) The energy-reaction graph for the reaction:
   $$A_2 + B_2 \rightleftarrows 2\ AB$$
   is shown at the right. Which of the following statements concerning that graph is *incorrect*?
   (a) The forward reaction is exothermic.
   (b) The energy of the reactants is higher than the energy of the products.
   (c) The activation energy is the energy difference between point 2 and point 1.
   (d) The potential energy of the products is higher than the potential energy of the reactants.

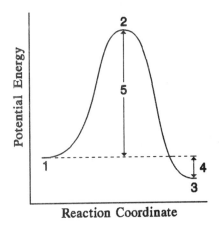

4) As temperature decreases, reaction rate decreases because:
   (a) the particles move too quickly for effective collisions to occur.
   (b) the kinetic energy of the particles becomes too high for bonds to form.
   (c) fewer particles have enough kinetic energy for an effective collision.
   (d) the orientation of the colliding particles becomes too random for an effective collision.

5) A catalyzed reaction occurs more quickly than an uncatalyzed reaction because:
   (a) the catalyst makes an alternative reaction pathway, with lower E_a available to the reactants.
   (b) the catalyst alters the kinetic energy of the reactants.
   (c) the catalyst changes the temperature of the mixture.
   (d) the catalyst makes the energy of reaction more negative.

6) As reactant concentration falls reaction rate falls because:
   (a) the particles move too slowly for an effective collision.
   (b) the activation energy increases.
   (c) the frequency of ineffective collisions increases.
   (d) the frequency of effective collisions decreases.

Assignment 19B: Le Chatelier's Principle; More Chemical Equilibria

In Assignment 19A you learned that temperature, catalysts and reactant concentration each affect reaction rate. It seems logical, then, to expect that if a system is at equilibrium, and we change one of these variables, the reverse reaction rates will no longer equal the forward reaction rates. What happens then? That is what you learn in this assignment.

Look for these big ideas:

1) According to **Le Chatelier's Principle**, if you do anything to alter the rates of reaction in an equilibrium, the equilibrium responds in a way that **counteracts partially the initial change** until a new equilibrium is reached.

2) A second way to interpret Le Chatelier's Principle might be: "Whatever a chemist does to an equilibrium system, the system itself tries to undo."

3) An equilibrium system can be described by an **equilibrium constant, K**. The constant K is a concentration ratio; the form of the ratio depends on how the equilibrium equation is written.

4) All equilibrium constants have the general algebraic form:

$$K = \frac{[\text{products}]}{[\text{reactants}]}$$

5) If K is very large, the equilibrium is favored in the forward direction; if K is very small, the equilibrium is favored in the reverse direction. If K is between 0.01 and 100, all reactants and products exist in appreciable concentrations at equilibrium.

Learning Procedures

Study
Sections 19.6–19.8, text pp. 516–523. Focus on your outline subheadings as you study.

Answer
Questions and Problems 25–58, text pp. 532–533. your answers with those on text p. A.41.

Take
the skills quiz below and on the next page. Check your answers with those on SG p. 172.

Assignment 19B Skills Quiz

Instructions: For Questions 1–4, select the letter of the *best* answer. Write the answers for Questions 5–7 in the spaces provided.

Questions 1–3 refer to the following equilibrium: $148 \text{ kJ} + A_2(g) + B_3(g) \rightleftarrows 2 \text{ AB}(g) + B(g)$

1) Removing some $B_3(g)$ causes the equilibrium to:
   (a) shift to left   (b) shift to right   (c) remain unchanged

2) Increasing the container volume causes the equilibrium to:
   (a) shift to left   (b) shift to right   (c) remain unchanged

3) Raising the temperature causes the equilibrium to:
   (a) shift to left   (b) shift to right   (c) remain unchanged

4) Vinegar is a 5% by weight solution of $HC_2H_3O_2$ in H_2O. The equation
$$HC_2H_3O_2(aq) + H_2O(\ell) \rightleftarrows H_3O^+(aq) + C_2H_3O_2^-(aq)$$
has an equilibrium constant $K_a = 1.8 \times 10^{-5}$. The substance present in greatest amount (after water) in vinegar is:
   (a) $H_3O^+(aq)$ only
   (b) $C_2H_3O_2^-(aq)$ only
   (c) $HC_2H_3O_2(aq)$ only
   (d) both $H_3O^+(aq)$ and $C_2H_3O_2^-(aq)$

Write equilibrium constant expressions for the following chemical equations.

5) $3 O_2(g) \rightleftarrows 2 O_3(g)$

6) $Al_2S_3(s) \rightleftarrows 2 Al^{3+}(aq) + 3 S^{2-}(aq)$

7) $H_3PO_4(aq) + H_2O(\ell) \rightleftarrows H_3O^+(aq) + H_2PO_4^-(aq)$

Assignment 19C: Equilibrium Calculations (optional)

This optional assignment explores the quantitative relationships between the size of the equilibrium constant for a reaction and the variable concentrations of reactants and products in that reaction. Look for the following big ideas:

1) All equilibrium calculations begin with the chemical equation for the reaction under study. Equilibria in water solutions are described using net ionic equations.

2) Given concentrations of substances in a reversible reaction, you can compute the value of the equilibrium constant for that reaction.

3) Given the initial concentrations of the substances that react to form an equilibrium and the numeric value of the equilibrium constant, you can determine the composition of the equilibrium mixture.

Learning Procedures

Study
Section 19.9, text pp. 523–530. Focus on your outline subheadings as you study.

Answer
Questions and Problems 59–78, text p. 533–534. Check your answers with those on text p. A.41.

Take
the skills quiz below and on the next page. Check your answers with those on SG pp. 172–173.

Assignment 19C Skills Quiz

1) Calculate the solubility of lead(II) iodide in moles per liter if $K_{sp} = 8.3 \times 10^{-9}$ for the equilibrium
$$PbI_2(s) \rightleftarrows Pb^{2+}(aq) + 2\,I^-(aq).$$

2) Find $[H^+]$, $[NO_2^-]$ and $[HNO_2]$ in a 1.1 M HNO_2 solution. The value of K_a is 4.5×10^{-4} for the equilibrium

$$HNO_2(aq) \rightleftarrows H^+(aq) + NO_2^-(aq).$$

3) How many grams of HNO_2 are needed along with 114 g KNO_2 to make a $[NO_2^-]/[HNO_2]$ buffer at pH = 3.50. The K_a for HNO_2 is given in Question 2 of this skills quiz.

4) At 225°C, 1.00 mol NOCl is placed in an empty 1.00 L vessel. The NOCl decomposes until the following equilibrium is established:

$$2\ NOCl(g) \rightleftarrows 2\ NO(g) + Cl_2(g)$$

At equilibrium, the vessel contains 0.045 mol $Cl_2(g)$. Calculate the value of K for this system at 225°C.

Answers to Chapter 19 Skills Quiz Questions

Assignment 19A
1) The answer is c. Read Section 19.1, text p. 509.
2) The answer is b. Study Section 19.2, text pp. 505–510, and also study Figure 19.1, text p. 510.
3) The answer is d. Check Section 19.3, text pp. 510–512, especially Figure 19.2 on text p. 511.
4) The answer is c. Study Section 19.4, text pp. 512–515 and Figure 19.4, p. 512.
5) The answer is a. See text pp. 513–514 and Figure 19.5, p. 513.
6) The answer is d. Study text pp. 514–515 and Figure 19.6, p. 514.

Assignment 19B
1) The answer is a. Study text pp. 516–517, top, particularly Examples 19.1 and 19.2, text p. 517.
2) The answer is b. Read text pp. 517–519, top. Study Examples 19.3–19.5.
3) The answer is b. See text p. 519. Study Examples 19.6–19.7 and take a long look at Figure 19.8, text p. 520. It's NO_2 formed by burning in the atmosphere that gives smog its characteristic brown color, which is always worse on hot summer days than cooler winter days.
4) The answer is c. Read text pp. 522–523; review text pp. 438–440. Pay particular attention to Quick Check 16.3 on p. 440.

For Questions 5–7, read Section 19.7, text pp. 520–522. Specific examples of each type of equilibrium constant expression are listed after each question.

5) $K = \dfrac{[O_3]^2}{[O_2]^3}$ See Example 19.8, text p. 521.

6) $K = [Al^{3+}]^2[S^{2-}]^3$ See Example 19.9 text p. 522.

7) $K = \dfrac{[H_3O^+][H_2PO_4^-]}{[H_3PO_4]}$ See Example 19.9 d, text p. 522.

Assignment 19C (optional)
1) $K_{sp} = [Pb^{2+}][I^-]^2 = 8.3 \times 10^{-9}$
   Let s = solubility, in moles per liter, of PbI_2, then $[Pb^{2+}] = s$ and $[I^-] = 2s$
   $(s)(2s)^2 = 8.3 \times 10^{-9}$ Note the parentheses in this step.
   $4s^3 = 8.3 \times 10^{-9}$
   $s = 1.3 \times 10^{-3}$ Study Example 19.12, text pp. 524–525.

2) $[H^+] = \sqrt{K_a[HA]} = \sqrt{(4.5 \times 10^{-4})(1.1)} = 0.022$ See Example 19.15, text p. 527.

3) At pH = 3.50, $[H^+] = 3.16 \times 10^{-4}$ (We'll round off once, at the end.)

   $4.5 \times 10^{-4} = \dfrac{[H^+][NO_2^-]}{[HNO_2]} = \dfrac{4.5 \times 10^{-4}}{3.16 \times 10^{-4}} = \dfrac{[NO_2^-]}{[HNO_2]} = 1.42$ continued top of next page...

$$114 \text{ g KNO}_2 \times \frac{1 \text{ mol KNO}_2}{85.1 \text{ g KNO}_2} \times \frac{1 \text{ mol HNO}_2}{1.42 \text{ mol KNO}_2} \times \frac{47.0 \text{ g HNO}_2}{1 \text{ mol HNO}_2} = 44 \text{ g HNO}_2$$

If you round 1.42 to 1.4 in the last step, you get 45 g HNO$_2$. See Example 19.17, text p. 528.

4) $2 \text{ NOCl(g)} \rightleftarrows 2 \text{ NO(g)} + \text{Cl}_2\text{(g)}$
I 1.00   0.   0.
R -0.090   +0.090 +0.045   Study Example 19.18, text pp. 493-494.
E 0.91   0.090   0.045

$$K = \frac{[NO]^2[Cl_2]}{[NOCl]^2} = \frac{(0.090)^2(0.045)}{(0.91)^2} = \frac{0.00037}{0.83} = 4.4 \times 10^{-4}$$

Sage Advice and Chapter Clues

Sometimes a picture is worth a thousand words. Look at the energy-reaction graph on the right. There's a lot of information there. The high activation energy E_a means the reaction rate is slow. The products are of higher energy than the reactants, so ΔE ($E_{products} - E_{reactants}$) is greater than zero, and this reaction is endothermic.

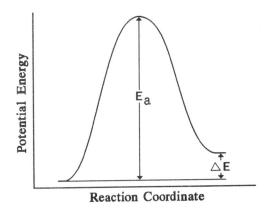

Anything you can do to ensure that more reactants have the needed E_a gives a faster reaction. If you raise temperature, all particles are moving faster, so more have the E_a needed for reaction. If you increase concentrations, a greater number of collisions occur. (Remember that the easiest way to increase the concentration of a gas is to reduce the volume.) If you add a catalyst, an additional reaction path with a lower E_a is available, so more particles react.

The major problem students have when writing equilibrium constant expressions is putting in too many terms. Pure solids and water do not appear in equilibrium expressions in water solution. The only concentrations listed in an equilibrium constant expression are those that are variable.

Let's use K_{sp} to illustrate the common problems students have with equilibrium. There are four trouble spots: 1) writing the chemical equation for the slightly soluble salt incorrectly; 2) including the concentration of the pure solid in the K_{sp} expression; 3) forgetting stoichiometry when setting up the solubility, s, terms; 4) forgetting parentheses with the solubility terms.

Bypass these trouble spots by remembering: 1) the K_{sp} expression demands that the solid appear on the left side of the chemical equation; 2) concentrations of pure solids are not included in equilibrium expressions; 3) if one mole of a salt dissolves to give two moles of a particular ion, that ion's concentration is 2s, not s; 4) (2s)^2 is not the same as 2s^2. See Question 1 in the Assignment 19C Skills Quiz for further details.

Weak acid equilibrium problems have one problem area, conversion between pH and [H$^+$]. This is usually an arithmetic problem, not a chemistry problem. If you are unsure how to manage these conversions on your calculator, find the calculator's instruction book. Still can't find that thing? You'll find the needed information in the text on p. A.1....

Chapter 19 Sample Test

Instructions: Circle the letter of the *best* answer for Questions 1–8. Answer Questions 9–13 in the spaces provided.

1) Which is the *incorrect* statement about equilibrium systems?
   (a) Forward and reverse reaction rates are the same at equilibrium.
   (b) The amounts of products and reactants are the same when the system reaches equilibrium.
   (c) No substance can enter or leave an equilibrium system.
   (d) Both chemical and physical equilibria exist.

2) Which is the *correct* statement about collision theory?
   (a) Most collisions are "effective" collisions.
   (b) Particles with low kinetic energy are most likely to have "effective" collisions.
   (c) Particles with high kinetic energy always have effective collisions.
   (d) Most collisions do not lead to a chemical reaction.

3) An energy-reaction graph is shown at the right. Which statement concerning the reaction on that graph is *correct*?
   (a) The products are of higher energy than the reactants.
   (b) The ΔE for the forward reaction is greater than zero.
   (c) The products are of lower energy than the reactants.
   (d) The activation energy E_a is the distance marked 2 on the graph.

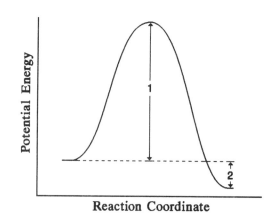

4) All of the following increase reaction rate *except*
   (a) lowering the temperature
   (b) adding a catalyst
   (c) raising concentrations
   (d) lowering the activation energy, E_a

Questions 5-7 refer to $2\,CO(g) + 2\,NO(g) \rightleftarrows 2\,CO_2(g) + N_2(g) + 747\,kJ$

5) Adding some NO(g) causes the equilibrium to
   (a) shift to left   (b) shift to right   (c) remain unchanged

6) Increasing the container volume causes the equilibrium to
   (a) shift to left   (b) shift to right   (c) remain unchanged

7) Lowering the temperature causes the equilibrium to
   (a) shift to left   (b) shift to right   (c) remain unchanged

8) The reaction $A(g) + B(g) \rightleftarrows AB(g)$ has an equilibrium constant K of 2.1×10^7. This reaction is favored
   (a) very strongly in the reverse direction.
   (b) slightly in the reverse direction.
   (c) very strongly in the forward direction.
   (d) slightly in the forward direction.

Write equilibrium constant expressions for the following reactions:

9) $2\,CO(g) + O_2(g) \rightleftarrows CO_2(g)$

10) $PbBr_2(s) \rightleftarrows Pb^{2+}(aq) + 2\,Br^-(aq)$

11) $NH_3(aq) + H_2O(\ell) \rightleftarrows NH_4^+(aq) + OH^-(aq)$

Check your sample test answers with those on SG p. 216.

Assignment 19C Sample Test (optional)

12) Calculate the solubility of silver sulfate, in moles per liter, if $K_{sp} = 1.2 \times 10^{-5}$.

13) What is the pH of a 1.2 M $HCHO_2$ solution if $K_a = 2.0 \times 10^{-4}$?

14) A solution contains 6.10 g $HCHO_2$ and 43.1 g $NaCHO_2$. At what pH is this solution buffered? K_a value given in Question 13, above.

15) At 325°C, the system $\qquad N_2(g) + 3\,H_2(g) \rightleftarrows 2\,NH_3(g)$ reaches equilibrium with $[N_2] = 0.057$, $[H_2] = 0.17$, $[NH_3] = 0.042$. Find K for the system at 325°C.

Check your sample test answers with those on SG pp. 216–217.

CHAPTER 20
Nuclear Chemistry

Assignment 20A: Natural Radioactivity. Where Does It Come From? Where Does It Go?

Most chemists study the results of electron sharing (covalent bonding, Chapters 10 and 11) and electron transfer (redox reactions, Chapter 18.) Some chemists and physicists, however, study the nucleus of the atom. In the nucleus, different "rules" of synthesis and decomposition apply. In this assignment, you will study some of those new rules.

The major ideas in this assignment are:

1) Three types of **natural radioactivity**, **alpha** (α), **beta** (β), and **gamma** (γ) rays can be formed when a nucleus decays. These rays differ in their masses, charges and their ability to penetrate matter.

2) Radioactivity can break chemical bonds by transferring enough energy to eject bonding electrons from atoms.

3) Radioactivity is detected by photographic film, scintillation counters, cloud chambers, or Geiger counters.

4) Each radioactive isotope possesses its own constant rate of decay.

5) The time needed for **one-half the radioactive atoms** in a sample to **decay** is the **half-life** of that sample. Half-life calculations can be done to determine how old a sample is, and how much is left at any time.

6) An equation for radioactive decay is balanced for nuclear charge (number of protons) and nuclear mass (number of protons and neutrons).

7) Products of radioactive decay may be other nuclei that undergo further decay, forming **a natural radioactive decay sequence**.

Learning Procedures
Study
Sections 20.1–20.6, text pp. 538–549. Focus on your outline subheadings as you study.

Answer
Questions and Problems 1–42, text pp. 559–560. Check your answers with those on text pp. A.42–43.

Take
the skills quiz on the next page. Check your answers with those on SG pp. 180–181.

Assignment 20A Skills Quiz

Instructions: Select the letter of the *best* answer to Questions 1–6. Solve Problem 7 in the space provided. You may use a "clean" periodic table.

1) Which radioactive emission is attracted by a + charge?
   (a) alpha ray   (b) beta ray   (c) gamma ray   (d) both b and c

2) Radioactivity is sometimes termed "ionizing radiation" because
   (a) all radioactive rays are ions.
   (b) radioactive rays remove energy from ions, when they collide.
   (c) radioactive rays remove positive ions from the air.
   (d) radioactive rays may form ions, when they collide with matter.

3) Which instrument uses a gas-filled tube to detect radioactivity?
   (a) photographic film   (b) scintillation counter   (c) cloud chamber   (d) Geiger counter

4) What fraction of a radioactive sample is left after the passage of two half-lives?
   (a) 1/8   (b) 1/4   (c) 1/2   (d) 1/1

5) The half-life of $^3_1 H$ is 12 years. If a sample contains 26 mg of $^3_1 H$ in 1994, how many mg of $^3_1 H$ will remain in that sample in 2059?

6) Using a Geiger counter, you find that the radioactivity of a radioactive isotope drops from 319 "units" to 25 "units" over 225 minutes. Use Figure 20.4, text p. 544, to determine the half-life of this isotope.

7) Balance the nuclear equation $^{240}_{93}Np \rightarrow\ ^{0}_{-1}e\ +$ _____.
   (a) $^{240}_{91}Pa$ (b) $^{239}_{92}U$ (c) $^{240}_{92}Np$ (d) $^{240}_{94}Pu$

8) A natural radioactive decay sequence always starts with a _____ isotope and ends with a _____ isotope.
   (a) radioactive, stable   (b) uranium, thorium   (c) stable, radioactive   (d) uranium, lead

Assignment 20B: Nuclear and Chemical Reactions, Induced Radioactivity, Uses of Radioactivity

We have been able to accomplish remarkable things through nuclear change, some good, some bad. We will look into a few of these now, using these ideas:

1) Nuclear reactions differ from ordinary chemical reactions because chemical reactions involve changes in valence electrons, but nuclear reactions involve changes in the nucleus.

2) **Nuclear bombardment reactions** make new isotopes and elements that do not now exist in nature. These isotopes and elements can produce "**induced radioactivity**."

3) The **transuranium elements**, whose atomic numbers are all **greater than 92**, are **all radioactive**.

4) A **nucleus that splits** into lighter nuclei undergoes **nuclear fission**.

5) A **chain reaction** occurs when a **product** of one reaction **is a reactant** in the next step of the reaction pathway.

6) Two light **nuclei that are joined** to form a heavier nucleus undergo **nuclear fusion**.

Learning Procedures

Study
Sections 20.7–20.13, text pp. 550–557, top. Focus on your outline subheadings as you study.

Answer
Questions and Problems 43–62, text pp. 560–561. Check your answers with those on text p. A.43.

Take
the skills quiz below and on the next page. Check your answers with those on SG p. 181.

Assignment 20B Skills Quiz

Instructions: Select the letter of the *best* choice.

1) Which statement is *correct*?
   (a) Radioactive $^{235}_{92}UF_6$ has different chemical reactions from $^{235}_{92}UF_6$.
   (b) The energy change in exothermic chemical reactions is about the same size as the energy change in nuclear reactions.
   (c) An isotope's oxidation number has no effect on its nuclear reactions.
   (d) An isotope's oxidation number has no effect on its chemical reactions.

2) Nuclear bombardment reactions are:
   (a) the end of the world.
   (b) sources of only nonradioactive isotopes.
   (c) collisions of particles that produce new isotopes.
   (d) sources of only radioactive particles.

3) Induced radioactivity comes from:
   (a) unstable isotopes that are heated to induce decomposition.
   (b) all isotopes produced in nuclear bombardment reactions.
   (c) background radiation.
   (d) radioisotopes produced in nuclear bombardment reactions.

4) Which of the following is a transuranium element?
   (a) $^{238}_{93}Np$ (b) $^{175}_{71}Yb$ (c) $^{238}_{92}U$ (d) $^{143}_{59}Pr$

5) Which of the following is a fission reaction?
   (a) $^{238}_{92}U + ^{1}_{0}n \rightarrow ^{239}_{92}U$
   (b) $^{235}_{92}U + ^{1}_{0}n \rightarrow ^{142}_{56}Ba + ^{91}_{36}Kr + 3\,^{1}_{0}n$
   (c) $^{13}_{6}C \rightarrow ^{13}_{7}N + ^{0}_{-1}e$
   (d) $^{9}_{4}Be + ^{4}_{2}He \rightarrow ^{12}_{6}C + ^{1}_{0}n$

6) Which of the equations in Question 5 describes a chain reaction?
   (a) a   (b) b   (c) c   (d) d

7) Which of the following is a fusion reaction?
   (a) $^{238}_{92}U + ^{1}_{0}n \rightarrow ^{239}_{92}U$
   (b) $^{6}_{3}Li + ^{1}_{0}n \rightarrow ^{3}_{1}H + ^{4}_{2}He$
   (c) $^{13}_{6}C \rightarrow ^{13}_{7}N + ^{0}_{-1}e$
   (d) $^{3}_{1}H + ^{2}_{1}H \rightarrow ^{4}_{2}He + ^{1}_{0}n$

Answers to Chapter 20 Skills Quiz Questions

Assignment 20A

1) Choice b is correct. Study Figure 20.1A, text p. 539.
2) Choice d is correct. Read text pp. 539–540.
3) Choice d is correct. See Section 20.3, text pp. 539–541, and study Figure 20.2 on page 540.
4) Choice b is correct. Review Equation 20.1, text p. 544, and study Figure 20.4 on page 544.

5) GIVEN: S = 26 mg $^{3}_{1}H$, 65 years, $t_{1/2}$ = 12 years   WANTED: (a) number of half-lives, n (b) mg $^{3}_{1}H$
   PATH: years → half-lives (a); half-lives → mg remaining (b)
   EQUATION: R = S × (0.5)^n (part b)
   (a) 65 years × $\frac{1 \text{ half-life}}{12 \text{ years}}$ = 5.42 half-lives;
   (b) R = 26 mg × (0.5)$^{5.42}$
       26 mg × 0.0234 = 0.61 mg

Read text pp. 544–547, especially Example 20.1 on pp. 544–545. If the exponential gave problems, check the example carefully then go to pp. A.4,5 in Appendix I for more calculator button pushing details.

6) GIVEN: S = 319, R = 25, time = 225 minutes, Figure 20.4, text p. 544   WANTED: $t_{1/2}$
   EQUATION: R/S = fraction of sample remaining
   $\frac{025}{319}$ = 0.078 of the sample remains.   From Fig. 20.4 this is 3.7 half-lives.

$$t_{1/2} = \frac{225 \text{ minutes}}{3.7 \text{ half-lives}} = 61 \text{ minutes}$$ See Example 20.3, text pp. 545–546.

7) Choice d is correct. Study Examples 20.5–20.6, text pp. 548–549.
8) Choice a is correct. Read text pp. 547–549.

Assignment 20B
1) Choice c is correct. Study text p. 550
2) Choice c is correct. Read Section 20.8, text pp. 550–552
3) Choice d is correct. See Section 20.8, especially page 551.
4) Choice a is correct. Review text p. 552.
5) Choice b is correct. Read text p. 553.
6) Choice b is correct. See text pp. 553–554, especially Figure 20.6, page 554.
8) Choice d is correct. Check text p. 556.

Sage Advice and Chapter Clues

You may think of nuclear chemistry as an untamed jungle, but there are rules to help you find the trails, just as you found the rules and trails in ordinary chemical reactions. For example, natural radioactivity has only three possible forms, as described below:

Emission	Mass (amu)	Charge	Penetration of Matter
alpha (α)	4	2+	extremely low
beta (β)	0	1-	low
gamma (γ)	0	0	very high

Ordinary chemical reactions depend on the valence electrons outside the nucleus; nuclear reactions depend on the protons and neutrons inside the nucleus. Changes in oxidation number or chemical bonding therefore have no effect on nuclear reactions, but a great effect on ordinary chemical reactions. Different isotopes of an element undergo vastly different nuclear reactions, but these isotopes all have very similar chemical reactivities.

Is nuclear chemistry a curse or a blessing? Like all new technology, it's both. The awesome destruction of atomic (fission) and hydrogen (fusion) weapons is only too well documented. While the nightmare of nuclear Armageddon is fading, the nightmare of nuclear proliferation becomes more real. (However, the "weapon of mass destruction" most easily built by a third-world country is a biological or chemical weapon, not a nuclear one.) The storage of radioactive waste, including deactivated nuclear warheads, remains an unsolved political problem. The health effects of radioactive waste are a concern.

But nuclear energy does not cause acid rain, black lung disease, or greenhouse effect, all undesirable byproducts of burning natural gas, fuel oil, coal or wood to heat homes or generate electricity. Ironically, nuclear power plants have by federal law lower radioactive emissions than coal or wood burning power plants!

The use of radionuclides by industry and medicine has made manufacturing more efficient, and therapy more healing. As an example of how nuclear chemistry affects you daily, consider the home smoke alarm. Battery powered smoke alarms depend on the radionuclide americium. No americium, no alarm.

So is nuclear chemistry a curse or a blessing? Like all tools, it depends how you wish to use it, so the answer depends on you.

Chapter 20 Sample Test

Instructions: Pick the letter of the *best* choice for Questions 1–4, 6–14; solve Question 5 in the space provided. You may use a "clean" periodic table.

1) Which natural radioactive emission is neither attracted nor repelled by either a + or a - charge?
   (a) alpha ray   (b) beta ray   (c) gamma ray   (d) both a and b

2) Ionizing radiation has an effect on body tissue because
   (a) the tissue gets a + charge.
   (c) physical changes occur.
   (b) the tissue gets a - charge.
   (d) chemical changes occur.

3) Geiger counter tubes operate when
   (a) the gas in the tube is ionized.
   (b) the gas in the tube fluoresces.
   (c) supersaturated vapor condenses on the ions in the tube.
   (d) a clicking sound is heard.

4) What fraction of a radioisotope is left after 3 half-lives have passed?
   (a) 1/8   (b) 1/9   (c) 1/3   (d) 1/6

5) In 24 hours, the mass of a radioisotope changed from 3.7 mg to 0.20 mg. Use Figure 20.4, text p. 508, to find the half-life of this isotope.

6) Balance the nuclear equation $^{221}_{87}\text{Fr} \rightarrow \underline{\qquad} + ^{4}_{2}\text{He}$.
   (a) $^{217}_{89}\text{Ac}$ (b) $^{217}_{85}\text{At}$ (c) $^{219}_{83}\text{Bi}$ (d) $^{225}_{89}\text{Ac}$

7) The three natural radioactive decay sequences all end with a _____ isotope as the final product.
   (a) $_{92}\text{U}$ (b) $_{90}\text{Th}$ (c) $_{82}\text{Pb}$ (d) $_{91}\text{Pa}$

8) Of the substances $^{239}_{94}\text{Pu}$, $^{239}_{94}\text{Pu}^{4+}$, $^{238}_{94}\text{Pu}$, $^{238}_{94}\text{Pu}^{4+}$ which two have the same chemical properties?
   (a) $^{239}_{94}\text{Pu}^{4+}$, $^{238}_{94}\text{Pu}$ (b) $^{239}_{94}\text{Pu}$, $^{238}_{94}\text{Pu}^{4+}$
   (c) $^{239}_{94}\text{Pu}^{4+}$, $^{238}_{94}\text{Pu}^{4+}$ (d) $^{239}_{94}\text{Pu}^{4+}$, $^{238}_{94}\text{Pu}$

9) Induced radioactivity comes from
   (a) stable products of bombardment reactions   (b) background radiation
   (c) radioactive products of bombardment reactions   (d) cosmic rays

10) Which is *not* a transuranium element?
    (a) $^{231}_{93}\text{Np}$ (b) $^{247}_{97}\text{Bk}$ (c) $^{244}_{94}\text{Pu}$ (d) $^{237}_{91}\text{Pa}$

The letters for the following reactions are the choices for 11–15.
   (a) $^{235}_{92}\text{U} + ^{1}_{0}\text{n} \rightarrow ^{144}_{54}\text{Xe} + ^{90}_{38}\text{Sr} + 2\,^{1}_{0}\text{n}$ (b) $^{96}_{42}\text{Mo} + ^{2}_{1}\text{H} \rightarrow ^{97}_{43}\text{Tc} + ^{1}_{0}\text{n}$
   (c) $^{3}_{2}\text{He} + ^{3}_{2}\text{He} \rightarrow ^{4}_{2}\text{He} + 2\,^{1}_{1}\text{H}$ (d) $^{20}_{8}\text{O} \rightarrow ^{20}_{9}\text{F} + ^{0}_{-1}\text{e}$

11) Which reaction above is a nuclear bombardment reaction?
    (a) a   (b) b   (c) c   (d) d

12) Which reaction above is a nuclear fission reaction?
    (a) a   (b) b   (c) c   (d) d

13) Which reaction above could be used in a nuclear chain reaction?
    (a) a   (b) b   (c) c   (d) d

14) Which reaction above is a nuclear fusion reaction?
    (a) a   (b) b   (c) c   (d) d

Check your sample test answers with those on SG p. 217.

CHAPTER 21

Organic Chemistry

Assignment 21A: Drawing and Naming Alkanes and Cycloalkanes

Over 11,000,000 compounds have been identified since 1965 in the chemistry laboratories of the world. Of these, about 95% are classed as organic compounds. This chapter gives you a brief survey of this important area of chemistry. The main ideas to be presented in this chapter are:

1) Because carbon atoms bond to each other forming chains, there is immense variety in carbon compounds.

2) Difficulties arise when we draw three dimensional structures on paper. Bending of a straight chain does *not* make a different compound.

3) **Hydrocarbons** are made of **carbon and hydrogen**. The **alkanes** are a hydrocarbon family with the general formula C_nH_{2n+2}, where n is the number of carbon atoms in the compound. The suffix **-ane** denotes an **alkane**; adding a prefix to the -ane suffix gives the number of carbon atoms in that straight chain alkane.

4) Compounds with the same molecular formula but different molecular structures are called **isomers**.

5) A **functional group** is an atom or group of atoms that determines both the identity of a class of compounds and its chemical properties.

6) Removing an H atom from an alkane gives an **alkyl functional group**. A general symbol for an alkyl residue, or radical, is **R**.

7) An alkane is named by using a prefix to indicate the number of carbon atoms in the longest chain. Numbers tell to which carbon atom(s) different functional groups are attached.

8) **Cycloalkanes** have all carbon-carbon single bonds, with at least some of the carbon atoms forming a ring.

9) Cycloalkanes are named according to the number of carbon atoms in the ring with the prefix *cyclo-*.

Learning Procedures

Study
Sections 21.1–21.3, text pp. 564–572. Focus on your outline subheadings as you study.

Answer
Questions and Problems 1–36, text pp. 599–601. Check your answers with those on text pp. A.43–44.

Take
the skills quiz on the next page. Check your answers with those on SG p. 190.

Assignment 21A Skills Quiz

Instructions: Pick the letter of the *best* choice for 1–4, 6; for 5 and 7, write the answer in the space given.

1) Which of these is considered to be organic?
   (a) CO_2
   (b) HCO_3^-
   (c) $CH_3CH_2CH_3$
   (d) CN^-

2) Which of the following is *not* an alkane?
   (a) CH_4
   (b) C_3H_6
   (c) $C_{10}H_{22}$
   (d) C_3H_8

3) C_3H_7- is a _____ group.
   (a) propyl
   (b) methyl
   (c) butyl
   (d) ethyl

4) Name the compound shown to the right:
   (a) 3,4-dimethylpentane
   (b) 1,2,3,4-tetramethylpropane
   (c) 2-ethyl-3-methylbutane
   (d) 2,3-dimethylpentane

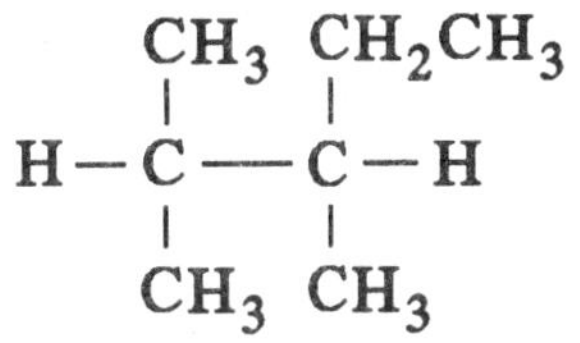

$$\begin{array}{cc} CH_3 & CH_2CH_3 \\ | & | \\ H-C-\!\!\!\!-C-H \\ | & | \\ CH_3 & CH_3 \end{array}$$

5) Draw the carbon skeleton of 2,3-dimethylbutane.

6) Name the compound shown to the right:
   (a) 1,3-dimethylcyclohexane
   (b) 2,4-dimethylbutane
   (c) 1,3-dimethylcyclobutane
   (d) 2,4-dimethylcyclobutane

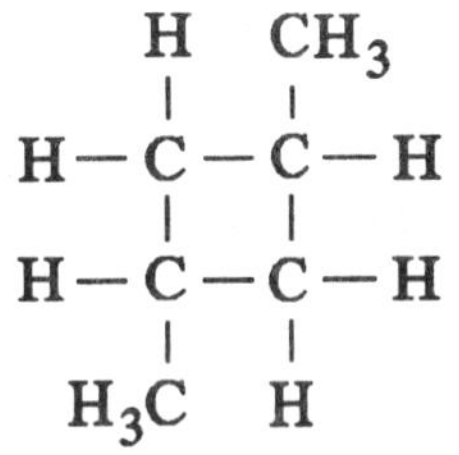

$$\begin{array}{cc} H & CH_3 \\ | & | \\ H-C-C-H \\ | & | \\ H-C-C-H \\ | & | \\ H_3C & H \end{array}$$

7) Draw the carbon skeleton of 1-bromo-2-methylcyclopentane.

Assignment 21B: More Hydrocarbons: Alkenes, Alkynes and Aromatics

The alkanes of Assignment 21A are called saturated, because they can form no additional bonds. The hydrocarbons in this assignment are unsaturated. You've probably already heard the words saturated and unsaturated used to describe fats in various foods. (Check the "Everyday Chemistry" on text page 597 for more about fats.) Look for these ideas:

1) The **alkenes** are hydrocarbons with at least one **double bond**; alkenes with one double bond have molecular formulas C_nH_{2n}. The suffix **-ene** denotes an alkene.

2) The **alkynes** are hydrocarbons with at least one **triple bond**; alkynes with one triple bond have

molecular formulas C_nH_{2n-2}. The suffix **-yne** denotes an alkyne.

3) In naming unsaturated compounds, the position of the double or triple bond is specified by number.

4) Double bonds can give geometric isomers, also called *cis, trans* isomers. Two alkyl groups can be on the same side (*cis*) or on opposite sides (*trans*) of the double bond.

5) **Benzene**, C_6H_6, is an **aromatic hydrocarbon**.

6) The carbon on a benzene ring to which a functional group is bonded is identified by number.

7) Most hydrocarbons are obtained by fractional distillation of crude oil.

8) Alkenes are made from **alcohols, R-OH**, or **alkyl halides, R-X**, where X = Br or Cl.

9) Multiple bonds in alkenes, but *not* in aromatic hydrocarbons, are opened by Cl_2, Br_2 or H_2 to give addition products.

10) Modern society depends upon hydrocarbons for fuel and as raw material for medicines, clothing fibers and plastics, and many other things we take for granted.

Learning Procedures
Study
Sections 21.4–21.9, text pp. 572–582. Focus on your outline subheadings as you study.

Answer
Questions and Problems 37–56, text pp. 601–602. Check your answers with those on text p. A.44.

Take
the skills quiz below and on the next page. Check your answers with those on SG p. 191.

Assignment 21B Skills Quiz

Instructions: Pick the letter of the *best* choice for 1, 4, 5; for 2 and 3, write the answer in the space given.

1) Name the compound on the right:
   (a) *cis*-3-pentene
   (b) *cis*-2-pentene
   (c) *trans*-3-pentene
   (d) *trans*-2-pentene

$$\begin{array}{cc} CH_3CH_2 & CH_3 \\ | & | \\ C = C \\ | & | \\ H & H \end{array}$$

2) Draw the carbon skeleton of 1-butyne.

3) Mothballs are 1,4-dichlorobenzene. Draw this compound, showing all the hydrogen atoms.

4) The product of the reaction at the right has the formula...
   (a) C₃H₆
   (b) C₃H₄
   (c) C₃H₅
   (d) C₃H₈

$$\begin{array}{c} CH_3 H \\ | | \\ C = C \\ | | \\ H H \end{array} + H_2 \xrightarrow{catalyst}$$

5) The product of the reaction to the right is:
   (a) 3,4-dibromohexane
   (b) *trans*-3-bromobutane
   (c) *cis*-3-bromobutane
   (d) no products are formed

$$\begin{array}{c} CH_3 CH_2CH_3 \\ | | \\ C = C \\ | | \\ CH_3CH_2 H \end{array} + Br_2 \longrightarrow$$

Assignment 21C: Organic Compounds with Oxygen or Nitrogen

The major organic functional groups include oxygen or nitrogen atoms. When these electronegative atoms are bonded to carbon, polar bonds are formed. These polar bonds may then be sites of chemical reactions. The main ideas are:

1) The general formula for an **alcohol** is **R-OH**; the suffix **-ol** denotes an alcohol. (Doesn't the model of ethanol on text p. 582 look like a puppy dog?)

2) Alcohols are usually made by addition of water to an alkene.

3) The general formula for an **ether** is **R-O-R'**; the word ether, preceded by the names of the attached alkyl groups is the name of the ether. Alcohols and ethers with the same number of carbon atoms are isomers.

4) **Aldehydes** and **ketones** have a **carbon atom double bonded to an oxygen atom**. In an aldehyde, the carbon is at the end of a chain; in a ketone, the carbon is inside a chain.

5) The IUPAC system uses the suffix **-al** for **aldehydes**, and the suffix **-one** for **ketones**. Ketones may also be named like ethers, but using the word ketone.

6) Aldehydes and ketones are prepared by oxidation of alcohols or hydration of alkynes. Aldehydes are themselves easily oxidized; ketones resist oxidation.

7) Aldehydes and ketones may be reduced to alcohols.

8) Oxidation of aldehydes, RCHO, gives **carboxylic acids, RCOOH**. IUPAC uses the suffix **-oic** and the word **acid** to denote carboxylic acids.

9) A **carboxylic acid and an alcohol** react to form an **ester**. Esters have two word names. The first word is the alkyl group from the alcohol, and the second is the anion derived from the acid. (Remember *-ic* → *-ate* in anions of acids.)

10) Organic molecules with nitrogen bonded to three atoms are **amines**. The IUPAC system names amines like ethers, but using the word amine.

11) All **amines** are Brönsted-Lowry and Lewis **bases**.

12) A **carboxylic acid and an amine** react to form an **amide**. Name the amide by replacing the -oic acid suffix with the word amide.

13) The amide group is found in the biological molecules called peptides and proteins.

Learning Procedures

Study

Sections 21.10–21.14, text pp. 582–592. Focus on your outline subheadings as you study.

Answer

Questions and Problems 57–80, text p. 602. Check your answers with those on text pp. A.44, 45.

Take

the skills quiz below and on the next page. Check your answers with those on SG p. 191.

Assignment 21C Skills Quiz

Instructions: For 1, 8 and 9, write the answers in the space provided. For 2–4, select the name of the *best* answer.

1) Write the Lewis diagrams for 1-pentanol, 2,2-dimethyl-1-propanol, and butyl methyl ether. What do all these substances have in common?

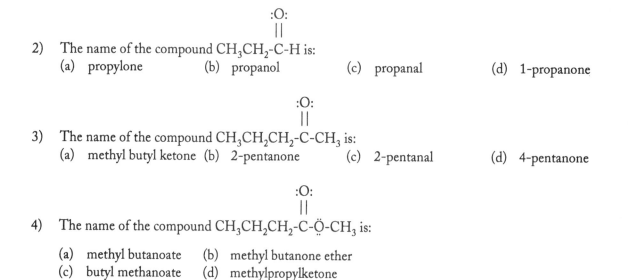

2) The name of the compound $CH_3CH_2\text{-}\overset{\overset{\displaystyle :O:}{\|}}{C}\text{-}H$ is:
 (a) propylone (b) propanol (c) propanal (d) 1-propanone

3) The name of the compound $CH_3CH_2CH_2\text{-}\overset{\overset{\displaystyle :O:}{\|}}{C}\text{-}CH_3$ is:
 (a) methyl butyl ketone (b) 2-pentanone (c) 2-pentanal (d) 4-pentanone

4) The name of the compound $CH_3CH_2CH_2\text{-}\overset{\overset{\displaystyle :O:}{\|}}{C}\text{-}\ddot{O}\text{-}CH_3$ is:

 (a) methyl butanoate (b) methyl butanone ether
 (c) butyl methanoate (d) methylpropylketone

5) Draw the Lewis diagram of dimethylamine.

6) Draw the Lewis diagram of propanamide.

Assignment 21D: Manufactured Polymers

Manufactured polymers, "plastics," are everywhere, and modern life would be impossible without them. Think not? Answer Question 1 in the skills quiz, and think again. Look for these big ideas concerning polymers.

1) Small molecules called **monomers** join together to form polymers. In analogy form, monomers are to links as polymers are to chains.

2) As molar mass of a polymer increases, intermolecular attractions (Section 14.2) increase, and mechanical strength increases.

3) **Addition polymers** are formed by repeated addition reactions of an alkene monomer to give an alkane-like polymer chain.

4) **Condensation polymers** are formed by repeated condensation reactions to give a polymer chain with repeated ester or amide functional groups.

5) Polymers that can be **melted and remolded** are called **thermoplastics**. Polymers that **harden irreversibly** upon heating are called **thermosets**.

6) Covalent bonds between polymer chains, called **cross-links** greatly increase polymer mass and **mechanical strength**.

Learning Procedures
Study
Sections 21.15–21.16, text pp. 592–596. Focus on your outline subheadings as you study.

Answer
Questions and Problems 81–94, text pp. 602–603. Check your answers with those on text p. A.45.

Take
the skills quiz below. Check your answers with those on SG pp. 191–192.

Assignment 21D Skills Quiz
1) Describe in as great a detail as possible how you awoke this morning and what you saw in your first waking moments.

2) Draw the structure of 1,1-dichloroethene. The addition polymer Velon™ is made from this monomer. Draw three repeating units of this polymer.

3) The structure of nylon 610 is shown below. Give the starting materials from which nylon 610 can be made by a condensation reaction.

$$\mathrm{{+}NH\text{-}CH_2\text{-}CH_2\text{-}CH_2\text{-}CH_2\text{-}CH_2\text{-}CH_2\text{-}NH\text{-}\underset{\overset{\|}{:O:}}{C}\text{-}CH_2\text{-}CH_2\text{-}CH_2\text{-}CH_2\text{-}CH_2\text{-}CH_2\text{-}CH_2\text{-}CH_2\text{-}\underset{\overset{\|}{:O:}}{C}{+}_n}$$

Answers to Chapter 21 Skills Quiz Questions

Assignment 21A

1) The correct answer is c. Read Section 21.1, text p. 564.
2) The correct answer is b. Check text pp. 566–568.
3) The correct answer is a. Study Table 21.2, text p. 566.
4) The correct answer is d. Read text pp. 568–569.
5) The carbon skeleton to the right is 2,3-dimethylbutane. Read text pp. 568–569.

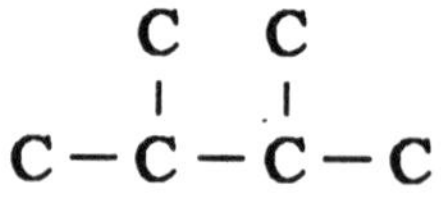

$$\begin{array}{c} C C \\ | | \\ C-C-C-C \end{array}$$

6) The correct answer is c. Read text pp. 571–572.

7) The molecule to the right is 1-bromo-2-methylcyclopentane.

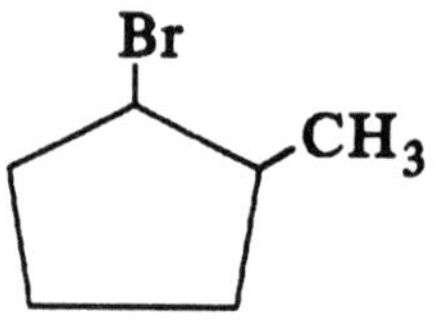

Assignment 21B

1) The correct answer is b. Read text p. 574.
2) The carbon skeleton of 1-butyne is C≡C—C—C. See text p. 574 for a discussion concerning 1-butene.
3) The molecule to the right is 1,4-dichlorobenzene. Study the text on p. 576.
4) The correct answer is d. Check text p. 578, top.
5) The correct answer is a. Read p. 579 in the text.

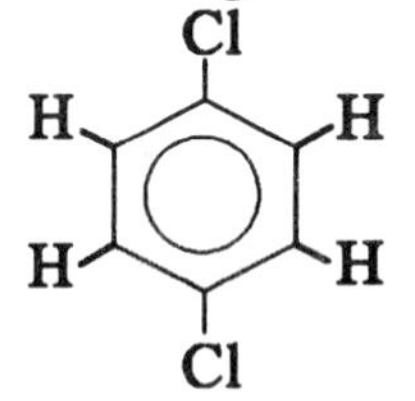

Assignment 21C

1)

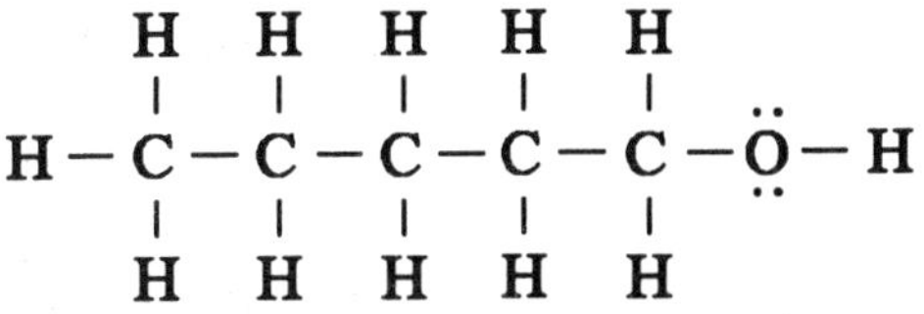

1-pentanol

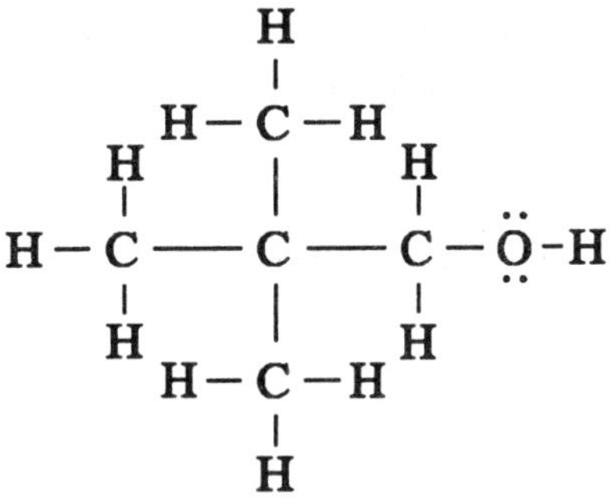

butyl methyl ether

2,2-dimethyl-1-propanol

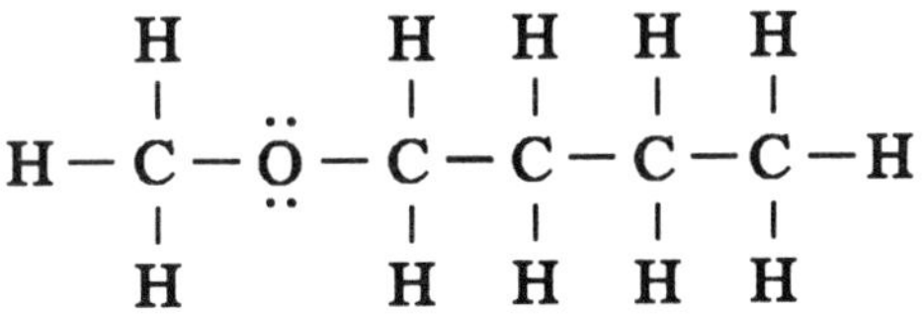

All these molecules are isomers having the formula $C_5H_{12}O$.
Read text pp. 582–584. Also see Problems 58 and 63, text p. 602.

2) The correct answer is c. See text pp. 585–587.
3) The correct answer is b. Read p. 586 in the text.
4) The correct answer is a. Check text pp. 587–589.

5) dimethylamine

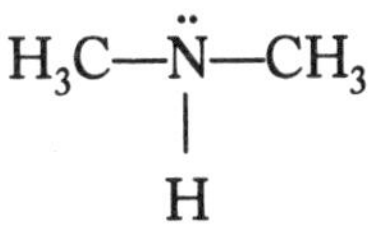

$H_3C—\ddot{N}—CH_3$
       |
       H

Read text p. 590.

6) propanamide

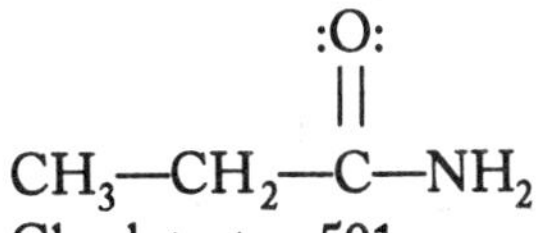

$CH_3—CH_2—\underset{\underset{\|}{:O:}}{C}—NH_2$

Check text p. 591.

Assignment 21D

1) Everyone's room is different, so we'll tell you what you *won't* see if you awoke in a polymerless room. The first commercial polymer, Celluloid, was patented in 1869. Bakelite, an early thermosetting polymer, was patented in 1909. Celluloid was used to make hair brush handles, toys, and as a substitute for amber, ivory and tortoise shell. Bakelite was typically used in buttons, drinking glasses, electrical plugs, sockets and switches, radio cabinets and control knobs, telephone bodies. If everything in your room was made before 1870 or so, there are probably no synthetic polymers in the room. Newer stuff than that? Let's remove all those polymers in your room, and see what's left...

First, how would you wake up? No alarm clocks (wind-up alarm clocks had Celluloid faceplates), no clock radios, no electricity. So you open your eyes. Can you see? No eyeglasses with plastic lenses or frames, no contact lenses. If you could focus, what would you see? The paint or wallpaper on the walls would be gone. Any upholstery and most carpeting would be gone, as would floor tiles or linoleum. Most bed linens and coverings would be gone. That room would be almost empty. Look closely, you'll be utterly amazed how little material would be left.

2)
$$\left[\begin{array}{cccccc} H & :\ddot{Cl}: & H & :\ddot{Cl}: & H & :\ddot{Cl}: \\ | & | & | & | & | & | \\ -C & -C & -C & -C & -C & -C- \\ | & | & | & | & | & | \\ H & :\ddot{Cl}: & H & :\ddot{Cl}: & H & :\ddot{Cl}: \end{array} \right]_n$$

Check text pp. 593–594, and Quick Check 21.9.

3) Two monomers from which nylon 610 could be made are $NH_2\text{-}CH_2\text{-}CH_2\text{-}CH_2\text{-}CH_2\text{-}CH_2\text{-}CH_2\text{-}NH_2$

and

$$HO\text{-}\overset{\overset{:\ddot{O}:}{\|}}{C}\text{-}CH_2\text{-}CH_2\text{-}CH_2\text{-}CH_2\text{-}CH_2\text{-}CH_2\text{-}CH_2\text{-}CH_2\text{-}\overset{\overset{:\ddot{O}:}{\|}}{C}\text{-}OH$$

Count the number of carbon atoms in the diamine to find the "6" in nylon 610; count the number of carbon atoms in the diacid to find the "10." Now read text pp. 595–596, and Question 94 on text p. 603.

Sage Advice and Chapter Clues

In Chapter 6, we urged you to *Learn the System!* when studying inorganic nomenclature. Memorize the minimum needed. In organic chemistry, that minimum is the list of functional groups. The functional group summary is Table 21.5 on page 592.

To what are functional groups attached? Hydrocarbons. The summaries of alkyl groups and hydrocarbons are Tables 21.2 (text p. 566), 21.3 (p. 573) and 21.4 (p. 577).

If you have trouble telling if two different drawings show the exact same molecule, or isomers, name both using the IUPAC rules. These rules force you to locate the longest carbon chain, regardless of how it's drawn. The two main IUPAC rules are: 1) find the longest carbon chain; 2) number the chain so the first branch on the chain has the lowest possible number.

You can remember the oxygen compounds as alkyl-group-substituted water molecules, and the amines as alkyl-group-substituted ammonia molecules:

water	H–O–H	ammonia	NH_3
alcohol	R–O–H	amines	RNH_2 or R_2NH or R_3N
ether	R–O–R'		

Primary alcohols and aldehydes are oxidized to carboxylic acids. Have you ever opened a bottle of spoiled wine? That wretched smell and taste were due to acetic acid, the final oxidation product of ethanol. (That's how wine vinegar is made.) If you were foolish enough to drink that brew, you probably got a dreadful hangover; the intermediate oxidation product acetaldehyde gives you that. Secondary alcohols oxidize to ketones, and no further.

As alcohols and acids react to form esters, acids and amines react to form amides. Because each of these reactions also gives one mole of water as a product, they are sometimes called condensation reactions. Some students have trouble spotting and naming esters. They see them and name them as a ketone next to an ether. They are not! They are esters, and they behave differently, and so we name them differently than ketones or ethers.

The easiest way to make a big molecule is to attach many little molecules to each other. That's the way polymers are made. Why make a polymer? because large molecules have different physical properties than little molecules. Let's use propane, $CH_3(CH_2)_1CH_3$, octane, $CH_3(CH_2)_6CH_3$, and eicosane, $CH_3(CH_2)_{18}CH_3$ as examples. All are nonpolar molecules with dispersion forces as their only attractive force. Propane is odorless (too small for the receptor sites in our noses), octane smells like, well, gasoline, and eicosane is also odorless (too large, too low a vapor pressure.) At room temperature, propane is a gas, octane a liquid, and eicosane a solid. Like all polymers, the larger the polymer, the higher the mechanical strength.

The easiest way for the chemist to increase polymer size is to add cross-links. If you have four molecules of molar mass 500,000, adding only three cross-links gives you one molecule of molar mass 2,000,000, with much higher mechanical strength. Read on for an familiar example of cross-linking.

Ever walk on a pair of stilts? Not an easy thing to do. An easier way to gain height is to add cross-links to the stilts to make a ladder. The ladder rungs are the cross-links. If you lean this ladder against a wall, you can climb it easily. Lean back too far, however, and you might tip the ladder backwards. The ladder rungs are cross-links in only two directions, horizontal and vertical. Take another cross-linked ladder, attach it to yours so the bottom of each ladder rests on the ground and the ladder tops are attached to each other, and *Voilà!* three-dimensional cross-links. (Most people call this more rigid, 3-dimensionally crosslinked polymer a step-ladder.)

One last thing. The first question in the Section 21D skills quiz wasn't whimsical. If you did wake up in a room surrounded completely by only natural fibers, with down for a comforter, wool for a blanket, cotton or silk for all clothing, you're surrounded by expensive materials. The chemical process industries have made modern life more democratic. While in the past, only the elite had a warm blanket, or a change of clean clothes, thanks to manufactured polymers, we take those comforts for granted today. Nylon may not be silk, and Comforel™ or Quallofil™ not goose down, but the manufactured products are a lot cheaper and more available; they're hypoallergenic, too....

Chapter 21 Sample Test

Instructions: Select the letter of the *best* choice for 1–3, 5–7, 10–13; for the rest, answer the question in the space provided.

1) The general formula C_nH_{2n} is the formula for the hydrocarbons called _____.
   (a) aliphatics (b) alkanes (c) alkynes (d) alkenes

2) The radical C_4H_9- depicts a _____ group.
   (a) hexyl (b) pentyl (c) butyl (d) propyl

3) The molecule to the right has the IUPAC name _____.
   (a) 1,1,2-trimethylpropane (b) 2-methylbutane
   (c) 1,2-dimethylbutane (d) 2,3-dimethylbutane

$$\begin{array}{cc} CH_3 & H \\ | & | \\ H-C\!\!-\!\!\!-\!\!\!-C-CH_3 \\ | & | \\ CH_3 & CH_3 \end{array}$$

4) Draw the carbon skeleton of 1,1-diethyl-3-methylcyclohexane.

5) Name the compound on the right:

(a) 1-ethyl-1-propene (b) *trans*-3-hexene
(c) *cis*-2-hexene (d) *cis*-3-hexene

$$\begin{array}{ccc} CH_3CH_2 & & H \\ | & & | \\ C & = & C \\ | & & | \\ H & & CH_2CH_3 \end{array}$$

6) The catalyzed reduction of an alkyne with one mole of hydrogen gas gives a(n) _____ hydrocarbon.
(a) alkane (b) alkene (c) cycloalkane (d) aromatic

7) Alkenes usually give _____ reactions; aromatic hydrocarbons like benzene usually give _____ reactions.
(a) addition; addition (b) substitution; addition (c) addition; substitution
(d) substitution; substitution

8) The aromatic hydrocarbon commonly called *mesitylene* is actually 1,3,5-trimethylbenzene. Draw the Lewis diagram for mesitylene.

9) Draw and give the IUPAC names for all the straight chain (no branches) alcohols having the formula $C_6H_{14}O$.

10) Primary alcohols can be oxidized to _____; secondary alcohols can be oxidized to _____.
   (a) aldehydes; carboxylic acids   (b) ketones; aldehydes   (c) aldehydes; carboxylic acids
   (d) aldehydes; ketones

11) The name of the compound $CH_3CH_2CH_2\text{-}\overset{\overset{\displaystyle :O:}{\|}}{C}\text{-}CH_2CH_3$ is:
   (a) 3-hexanone   (b) butyl ethyl ketone   (c) 3-hexanal   (d) 4-hexanone

12) The name of the compound $CH_3CH_2CH_2\text{-}\overset{\overset{\displaystyle :O:}{\|}}{C}\text{-}\ddot{O}\text{-}CH_2CH_3$ is _____, although it's called "pineapple oil" in the flavor and fragrance industry.
   (a) butyl ethanoate   (b) butyl acetate   (c) butyl ethanone   (d) ethyl butanoate

13) The condensation reaction between a carboxylic acid and an amine gives an _____.
   (a) ester   (b) anhydride   (c) amide   (d) carboxylic acid amine

14) The addition polymer Kel-F is formed from chlrotrifluoroethene. Draw three repeating units of Kel-F.

15) The addition polymer shown at the right is commonly called povidone, and is used in eye drops as a lubricant. Give the monomer from which povidone can be made.

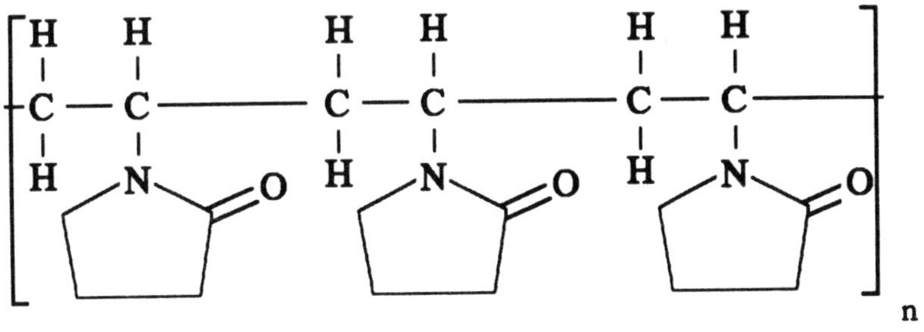

Check your sample test answers with those on SG pp. 217–218.

CHAPTER 22

Biochemistry

Assignment 22A: Amino Acids and Proteins

Amino acids are small molecules with 3 functional groups. One functional group is always an amine; a second functional group is always a carboxylic acid. The third functional group, designated an R group, determines the identity of the amino acid. The definition of an R group has been expanded from Section 21.3 to mean *any* group of atoms, not just an alkyl group, that completes a Lewis diagram.

Amino acids bond with each other by forming amide bonds. As the length of the amino acid chain grows, we obtain peptides, and then proteins.

Macromolecules such as proteins can have varying shapes in different parts of the molecule. To describe protein shape we need to know its primary, secondary and tertiary structures. Protein function can only be understood in terms of protein structure.

Look for these big ideas:

1) There are 20 "standard" amino acids; each amino acid is known by its name, or by a 3 letter or a 1 letter code.

2) Proteins are polymers made from amino acid monomers. With 20 monomers, the number of possible proteins is vast, estimated to be about 8,000 in a fruitfly, about 100,000 in us.

3) **Primary structure** is the **Lewis diagram**, or the amino acid codes in correct order, of a protein.

4) **Secondary structure** is the *local* spatial layout of the amino acid backbones. Two secondary structures are the α-**helix** and the β-**pleated sheet**. Secondary structures allow for maximum hydrogen bonding (Section 14.2), and greatest stability.

5) **Tertiary structure** describes the spatial arrangement of the *entire* peptide or protein molecule.

6) **Fibrous proteins** form **connective tissues** such as hair, nails, or collagen; fibrous proteins have a rod-like extended shape and are insoluble in water. **Globular proteins** have a highly folded spherical shape and are water soluble.

7) **Enzymes** are globular proteins that serve as catalysts to control the rate of specific chemical reactions in a living system. An enzyme **substrate** is the starting material in the catalyzed reaction. The **lock and key analogy** is a model for enzyme-substrate interaction.

8) An enzyme's substrate binds at the enzyme's **active site**. Enzyme **inhibitors** compete with substrate for the active site. Inhibitors are reversible or irreversible.

Learning Procedures

Study
Sections 22.1–22.2, text pp. 608–615. Focus on your outline subheadings as you study.

Answer
Questions and Problems 1–20, text p. 629. Check your answers with those on text pp. A.46–47.

Take
the skills quiz below. Check your answers with those on SG p. 202.

Assignment 22A Skills Quiz

Instructions: Answer 1 in the space provided. For 2–6, select the letter of the *best* choice answer. You may use Table 22.1, text p. 609 for amino acid structures and abbreviations.

1) The artificial sweetener aspartame, tradename NutraSweet®, is the methyl ester of the dipeptide aspartyl-phenylalanine, abbreviated Asp-Phe, or D-F. Draw the structure of aspartylphenylalanine.

2) How many different tripeptides can be made from one glycine, G, one tryptophan, W, and serine, S?
 (a) 2 (b) 3 (c) 6 (d) 8 (e) 9

3) The primary structure of a protein refers to its:
 (a) hydrogen bonding within a given chain
 (b) sequence of amino acids
 (c) helical structure
 (d) disulfide linkages

4) The secondary structure of a protein is derived from:
 (a) peptide linkages
 (b) disulfide linkages
 (c) hydrogen bond formation
 (d) sequence of amino acids

5) Hydrogen bonding in a β-pleated sheet secondary structure is best described as:
 (a) at a minimum, and occurring between amino acids in the same chain
 (b) at a maximum, and occurring between amino acids in the same chain
 (c) at a minimum, and occurring between amino acids in adjacent chains
 (d) at a maximum, and occurring between amino acids in adjacent chains

6) Zestril® is a synthetic dipeptide that lowers blood pressure by inhibiting the Angiotensin Converting Enzyme. Called an ACE inhibitor, Zestril® is taken once a day and is excreted unchanged in the urine. Zestril is a(n) _____ inhibitor of ACE.
 (a) reversible
 (b) irreversible
 (c) both reversible and irreversible
 (d) neither reversible nor irreversible

Assignment 22B: Carbohydrates

Carbohydrates, or as the French say "hydrates of carbon," with empirical formula $[C \cdot (H_2O)_n]$, are the most important fuel molecules for organisms. Due to the abundance of the polymers cellulose and starch, there is probably more carbohydrate in the biosphere than all other organic matter combined.

We'll start small, and work our way up to polymers. Look for these big ideas as you study:

1) The most important **monosaccharide** (one sugar) is **glucose**; monosaccharides are also called simple sugars.

2) Monosaccharides like glucose have both open chain and cyclic forms, with the cyclic forms as the major form. There are two isomers, called anomers, of monosaccharides like glucose that vary only at carbon 1.

3) The form called β-glucose is the chemically most stable form, and the monomer for many larger polymeric carbohydrates.

4) **Disaccharides** are formed from two simple sugars. Important disaccharides are sucrose (table sugar), a glucose-fructose compound, and lactose (milk sugar), a galactose-glucose compound.

5) **Polysaccharides** (many sugars) serve both structural and energy storage roles. Cellulose, starch, amylopectin and glycogen are examples of polysaccharides.

6) Like manufactured polymers, as cross-linking in a polysaccharide increases, so does molar mass and mechanical strength. As molar mass increases, water solubility decreases.

Learning Procedures

Study
Section 22.3, text pp. 615–620. Focus on your outline subheadings as you study.

Answer
Questions and Problems 21–34, text pp. 629–630. Check your answer with those on text p. A.47.

Take
the skills quiz below. Check your answers with those on SG p. 202.

Assignment 22B Skills Quiz

Instructions: Select the letter of the *best* choice answer.

1) Which 2 monosaccharides below are anomers?

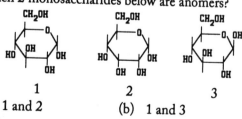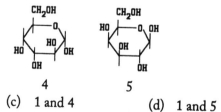

    (a) 1 and 2   (b) 1 and 3   (c) 1 and 4   (d) 1 and 5

2) What are the monosaccharide units in sucrose?
    (a) glucose   (b) galactose   (c) fructose   (d) a and b   (e) a and c

3) Which disaccharide is found in milk?
   (a) sucrose (b) lactose (c) fructose (d) amylose

4) Carbohydrates are stored in mammals as...
   (a) cellulose (b) amylopectin (c) glucose (d) glycogen

5) Which monosaccharide is obtained when glycogen is hydrolyzed (broken into monosaccharides)?
   (a) galactose (b) glucose (c) fructose (d) ribose

Assignment 22C: Lipids

"Oil and water don't mix" is a common saying, because it's true. It's also true in living systems. The general term *lipid* describes a substance that doesn't dissolve in water, but does dissolve in nonpolar solvents like hexane or ether.

Although we think of ourselves as water-based creatures, lipids play vital roles in our lives. There is no life without the cell; there is no cell without the cell membrane; there is no cell membrane without lipids. You often hear terms such as saturated and unsaturated fats, triglycerides (now called triacylglycerols); anabolic steroids ('roids); birth control pills, cholesterol, cortisone; all these are lipids found in this assignment.

Look for these big ideas:

1) **Fats and oils** are esters of long chain acids and glycerol, 1,2,3-propanetriol. Most fats come from animal products; most oils come from plant products.

2) At room temperature, fats are solids and oils liquids. Saturated fats have higher melting points than unsaturated fats.

3) **Phospholipids** are also esters of glycerol, but one acid is phosphoric acid, not a long chain carboxylic acid. Phospholipids are necessary for cell membranes.

4) **Waxes** are esters made from long chain carboxylic acids and long chain alcohols.

5) **Steroids** have the four ring framework shown below:

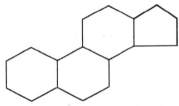

The major steroid in animals is cholesterol, a necessary structural material. A 150 lb person contains about 200 g of cholesterol, as both the free alcohol and fatty acid esters.

Learning Procedures

Study
Section 22.4, text pp. 620–624. Focus on your outline subheadings as you study.

Answer
Questions and Problems 35–42, text p. 630. Check your answers with those on text p. A. 47.

Take
the skills quiz on the next page. Check your answers with those on SG p. 202.

Assignment 22C Skills Quiz

Instructions: Pick the letter of the *best* choice answer.

1) Which of the following statement regarding lipids is *not* true?
   (a) Many lipids have biological roles.
   (b) All lipids have the same functional groups.
   (c) Lipids are soluble in nonpolar solvents.
   (d) Lipids include triacylglycerols, waxes and steroids.

2) Which of the following substances is a lipid?
   (a) alanine   (b) glucose   (c) fat   (d) glycogen

3) Which of these is a wax?

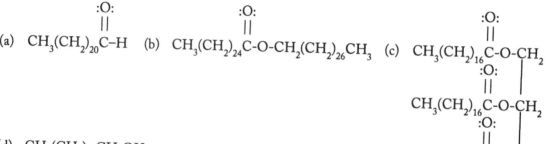

   (d) $CH_3(CH_2)_{20}CH_2OH$

4) Which of these is a human sex hormone?
   (a) testosterone   (b) cortisone   (c) cholesterol   (d) lecithin

---

Assignment 22D: Nucleic Acids

Why do children resemble their parents? What process enables transfer of genetic information from parents to children? What molecules are involved in this process? Important questions, important molecules. Look for these big ideas:

1) Two types of **nucleic acid polymers** are **DNA**, deoxyribonucleic acid and **RNA**, ribonucleic acid.

2) Nucleic acid **monomers** are called **nucleotides**. Each nucleotide is composed of a cyclic nitrogen compound called a base, a sugar, and a phosphate group. The nucleotide sequence is the primary structure of a nucleic acid.

3) There are 5 nucleic acid nitrogen bases: thymine, (abbreviated T), cytosine, C, adenine, A, guanine, G, and uracil, U. Thymine is found only in DNA; uracil is found only in RNA.

4) The **secondary structure** of **DNA** is a **double helix**; the secondary structure of **RNA** is generally a **single helix**.

5) The DNA **double helix** is stabilized by **interchain hydrogen bonding** between nitrogen bases. Bases that hydrogen bond with each other are called **complementary base** pairs, or complements.

6) In DNA, adenine, A, forms hydrogen bonds to thymine, T; guanine, G, forms hydrogen bonds to cytosine, C.

7) In **replication**, a DNA molecule duplicates itself in a fashion dubbed "semi-conservative."

8) In **transcription**, an RNA molecule complementary to a DNA strand is assembled. This messenger RNA (mRNA) then travels out of the cell nucleus. In mRNA only, uracil, U, is the base complementary to the DNA's adenine.

9) In **translation**, the mRNA is decoded and individual amino acids are brought by different transfer RNA (tRNA) molecules to be assembled into proteins.

Learning Procedures

Study
Section 22.7, text pp. 624–627. Focus on your outline subheadings.

Answer
Questions and Problems 43–58, text p. 630. Check your answers with those on text pp. A. 47–48.

Take
the skills quiz below. Check your answers with those on SG p. 202.

Assignment 22D Skills Quiz

Instructions: Select the letter of the *best* choice answer.

1) The nitrogen base found *only* in RNA is
   (a) uracil (b) thymine (c) guanine (d) cytosine

2) A nucleotide is composed of:
   (a) a 5 carbon sugar (b) a phosphate (c) a nitrogen base (d) an amino acid (e) a, b and c

3) In a nucleotide the components are, in order, _____.
   (a) monosaccharide - phosphate - nitrogen base (b) amino acid - monosaccharide - phosphate
   (c) phosphate - monosaccharide - nitrogen base (d) nitrogen base - phosphate - monosaccharide

4) Which bases pair in DNA by hydrogen bonding?
   (a) cytosine, thymine (C,T) (b) cytosine, uracil (C,U)
   (c) adenine, guanine (A,G) (d) adenine, thymine (A,T)

5) Which term describes the synthesis of a new DNA molecule that is a copy of an existing DNA molecule?
   (a) duplication (b) transcription (c) translation (d) replication

6) If the base sequence along a segment of DNA were A-T-G, what would be the complementary base sequence?
   (a) A-A-G (b) C-T-G (c) T-A-C (d) C-A-T

7) If the base sequence along a segment of DNA were T-A-C, what would be the base sequence of the messenger RNA synthesized from this sequence?
   (a) C-G-T (b) A-U-G (c) T-A-C (d) U-A-C

Answers to Chapter 22 Skills Quiz Questions

Assignment 22A

1) Aspartylphenylalanine is to the right. See text pp. 608–609,

   NutraSweet has the COOH group on the far right changed to a methyl ester.

   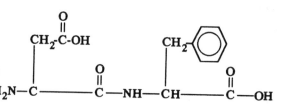

2) The answer is c. Check text p. 610, particularly Quick Check 22.1.
3) The answer is b. Back to text p. 610.
4) The answer is c. To text pp. 611–612 with you.
5) The answer is d. Read text pp. 611–612.
6) The answer is a. See text pp. 613–615, especially Quick Check 22.3.

Assignment 22B

1) The answer is d. See text p. 616.
2) The answer is e. Check p. 617 in the text.
3) The answer is b. Read p. 618 in the text.
4) The answer is d. Study p. 620 in the text.
5) The answer is b. Back to p. 620 in the text.

Assignment 22C

1) The answer is b. Check p. 620 in the text.
2) The answer is c. You must study the whole section, pp. 620–624, to put the pieces together for this one.
3) The answer is b. Read text p. 622.
4) The answer is a. Go back to Figure 22.4, text p. 623.

Assignment 22D

1) The answer is a. Study Figure 22.5, text p. 624.
2) The answer is e. Study text pp. 624–625.
3) The answer is c. See Figure 22.6 on text p. 625 and Quick Check 22.6 on p. 627 in the text.
4) The answer is d. Read page 625 in the text. (*There! I've done it! I've spelt the word page out once!*)
5) The answer is d. Check text p. 626.
6) The answer is c. Figures 22.7, 22.8, text pp. 626, 627, and Questions 53 and 54, text p. 630.
7) The answer is b. Check text p. 626 and study Questions 55 and 56, text p. 630.

Chapter 22 Sample Test

Instructions: Select the *best* answer for each question.

1) How many different tripeptides can be made from two valines, V and one leucine, L?
   (a) 2   (b) 3   (c) 6   (d) 8   (e) 9

2) Polymers of amino acids are called _____.
   (a) polysaccharides   (b) nucleic acids   (c) fats and oils   (d) proteins

3) Hydrogen bonding in an α-helix protein secondary structure is best described as:
   (a) at a minimum, and occurring between amino acids in the same chain
   (b) at a maximum, and occurring between amino acids in the same chain
   (c) at a minimum, and occurring between amino acids in adjacent chains
   (d) at a maximum, and occurring between amino acids in adjacent chains

4) Enzymes are _____.
   (a) carbohydrates   (b) fats   (c) lipids   (d) proteins   (e) DNA

5) Which monosaccharide below is β-glucose?

   (a) 1   (b) 2   (c) 3   (d) 4   (e) 5

6) What are the monosaccharide units in lactose?
   (a) glucose   (b) galactose   (c) fructose   (d) a and b   (e) a and c

7) The glucose polymer responsible for the rigidity of a plant stalk is _____.
   (a) starch   (b) amylopectin   (c) glycogen   (d) amylose   (e) cellulose

8) Which functional group is found in all fats?
   (a) ketone   (b) ester   (c) alkene   (d) acid   (e) alcohol

9) Which statement concerning fats and oils is *incorrect*?
   (a) The greater the unsaturation, the higher the melting point?
   (b) Solid examples are called "fats."
   (c) Such compounds are insoluble in water.
   (d) These compounds are also called triacylglycerols.

10) Which of these is a triacylglycerol (triglyceride)?

(a) $CH_3(CH_2)_{20}\overset{\overset{\displaystyle :O:}{\|}}{C}-H$

(b) $CH_3(CH_2)_{24}\overset{\overset{\displaystyle :O:}{\|}}{C}-O-CH_2(CH_2)_{26}CH_3$

(c) $CH_3(CH_2)_{16}\overset{\overset{\displaystyle :O:}{\|}}{C}-O-CH_2$
$CH_3(CH_2)_{16}\overset{\overset{\displaystyle :O:}{\|}}{C}-O-CH_2$
$CH_3(CH_2)_{16}\overset{\overset{\displaystyle :O:}{\|}}{C}-O-CH_2$

(d) $CH_3(CH_2)_{20}CH_2OH$

11) Which of these is a human sex hormone?
 (a) lecithin
 (b) cholesterol
 (c) cortisone
 (d) progesterone

12) The nitrogen base found *only* in DNA is
 (a) uracil
 (b) thymine
 (c) guanine
 (d) cytosine

13) Which term describes the synthesis of a protein in a living system?
 (a) duplication
 (b) transcription
 (c) translation
 (d) replication

14) If the base sequence along a segment of DNA were G-A-C, what would be the complementary base sequence?
 (a) C-T-G
 (b) T-C-A
 (c) C-A-G
 (d) G-T-C

Check your sample test answers with those on SG p. 218.

Answers to Sample Test Questions

Chapter 2

1) The answer is c.
2) The answer is d.
3) The answer is c.
4) The answer is a.
5) The answer is d.
6) The answer is d.
7) The answer is c.
8) The answer is below.
9) The answer is d.
10) The answer is a.
11) The answer is a.
12) The answer is c.
13) The answer is b.

8) The reactants are hydrogen, H_2 and iodine, I_2; the product is hydrogen iodide, HI. The H_2 and the I_2 are both elements; the HI is a compound.

Chapter 3

1) $413,400 = 4.13400 \times 10^5$ $\qquad$ $6.91 \times 10^7 = 69,100,000$
   $0.00103 = 1.03 \times 10^{-3}$ $\qquad$ $1.47 \times 10^{-4} = 0.000147$

2) $4.1 \times 10^{-6} + 1.59 \times 10^{-5} = 2.00 \times 10^{-5}$ $\qquad$ $6.7 \times 10^3 + 2.61 \times 10^4 = 3.28 \times 10^4$
   $7.14 \times 10^3 - 3.9 \times 10^2 = 6.75 \times 10^3$ $\qquad$ $8.34 \times 10^{-1} - 3.6 \times 10^{-2} = 7.98 \times 10^{-1}$

3) $(1.16 \times 10^{-3})(6.32 \times 10^{-11}) = 7.33 \times 10^{-14}$
   $(4.62 \times 10^{-6})(2.17 \times 10^8) = 1.00 \times 10^3$
   $(5.71 \times 10^4)(7.45 \times 10^{14}) = 4.25 \times 10^{19}$
   $\dfrac{(9.76 \times 10^{-7})(8.17 \times 10^3)}{(1.23 \times 10^{-1})} = 6.48 \times 10^{-2}$ $\qquad$ $\dfrac{-4.39 \times 10^4}{(107)(7.11 \times 10^1)} = -5.77$

4) At the Lagrange points a 72 kg person still has a mass of 72 kg, although this person's weight would be zero.

5) The prefix μ means 10^{-6}, so there are 10^{-6} phones in 1 μphone; the prefix M means 10^{+6}, so there are 10^{+6} phones in 1 Mphone. There must then be 10^{+12} μphones in 1 Mphone.

6) The temperature 0°F is colder than 0°C.

7) The measurement 0.099 g has two significant figures.

8) To three significant figures 2.6034 km rounds to 2.60 km.

9) The answer is 145.92.

10) The answer rounds to 0.0027.

11) $3 \text{ months} \times \dfrac{4.33 \text{ week}}{1 \text{ month}} \times \dfrac{13.6 \text{ hr}}{1 \text{ week}} \times \dfrac{\$6.45}{1 \text{ hr}} = \$1139.48 \rightarrow \$1139$

12) $45.0 \text{ in} \times \dfrac{2.54 \text{ cm}}{1 \text{ in}} = 114 \text{ cm}$

13) $40.1 \text{ m} \times \dfrac{1000 \text{ mm}}{1 \text{ m}} = 40,100 \text{ m} = 4.01 \times 10^4 \text{ m}$

14) $9.45 \text{ km} \times \dfrac{1000 \text{ m}}{1 \text{ km}} = 9450 \text{ m}$

15) $3440 \text{ cm}^3 \times \dfrac{1 \text{ mL}}{1 \text{ cm}^3} \times \dfrac{1 \text{ L}}{1000 \text{ mL}} \times \dfrac{1.06 \text{ qt}}{1 \text{ L}} = 3.65 \text{ qt}$

16) $439 \text{ cg} \times \dfrac{1 \text{ g}}{100 \text{ cg}} \times \dfrac{1 \text{ oz}}{28.3 \text{ g}} = 0.155 \text{ oz}$

17) 14 - 32 = (1.8°C)
    °C = (14 - 32)/1.8 = (-18)/1.8
    °C = -10

18) °F - 32 = 1.8(71)
    °F - 32 = 128
    °F = 160

19) $T_K = T°_C + 273.15 = 312 + 273.15 = 585$ K

20) (a) GIVEN: 377.03 g full mass, 15.863 g empty mass, 354 mL volume   WANTED: density

    EQUATION: $D = \dfrac{m}{V} = \dfrac{377.03 \text{ g} - 15.863 \text{ g}}{354 \text{ mL}} = 1.02$ g/mL

    (b) The density of the water in the fish tank must be between 1.02 g/mL and 1.05 g/mL.

Chapter 4

1) The answer is b.   2) The answer is b.

3)

	Volume	Temperature	Pressure	Amount
Initial Value (1)	0.610 L	constant	0.103 atm	constant
Final Value (2)	V_2	constant	1.62 atm	constant

$0.610 \text{ L} \times \dfrac{0.103 \text{ atm}}{1.62 \text{ atm}} = 0.0388$ L

4)

	Volume	Temperature	Pressure	Amount
Initial Value (1)	constant	-1°C = 272 K	462 torr	constant
Final Value (2)	constant	T_2	591 torr	constant

$T_2 = \dfrac{591 \text{ torr} \times 272 \text{ K}}{462 \text{ torr}} = 348$ K   °C = K - 273 = (348 - 273) = 75°C

5)

	Volume	Temperature	Pressure	Amount
Initial Value (1)	2.14 L	40°C = 313 K	constant	constant
Final Value (2)	V_2	20°C = 293 K	constant	constant

$2.14 \text{ L} \times \dfrac{293 \text{ K}}{313 \text{ K}} = 2.00$ L

6)

	Volume	Temperature	Pressure	Amount
Initial Value (1)	4.80 L	32°C = 305 K	744 torr	constant
Final Value (2)	V_2	64°C = 337 K	810 torr	constant

$4.80 \text{ L} \times \dfrac{337 \text{ K}}{305 \text{ K}} \times \dfrac{744 \text{ torr}}{810 \text{ torr}} = 4.87$ L

7)

	Volume	Temperature	Pressure	Amount
Initial Value (1)	1.24 L	0°C = 273 K	1.00 atm	constant
Final Value (2)	V_2	21°C = 294 K	1.21 atm	constant

Continued on next page.

$$1.24 \text{ L} \times \frac{294 \text{ K}}{273 \text{ K}} \times \frac{1.00 \text{ atm}}{1.21 \text{ atm}} = 1.10 \text{ L}$$

Chapter 5

1) The answer is b. 2) The answer is c. 3) The answer is d. 4) The answer is a.
5) The answer is d. 6) The answer is b. 7) The answer is b. 8) The answer is b.
9) The answer is d. 10) The answer is a.
11)
Br   bromine   sodium   Na
Mg   magnesium   nickel   Ni
Pb   lead   phosphorus P
Fe   iron   calcium   Ca
Ag   silver   silicon   Si
K   potassium   fluorine   F
Hg   mercury   manganese   Mn

Chapter 6

bromine	Br_2	H_2	hydrogen
nitrogen	N_2	O_2	oxygen
chlorine	Cl_2	F_2	fluorine
magnesium oxide	MgO	CaI_2	calcium iodide
sodium fluoride	NaF	Ba^{2+}	barium(II) ion
aluminum nitride	AlN	Na_2S	sodium sulfide
calcium phosphide	Ca_3P_2	Cl^-	chloride ion
potassium bromide	KBr	Li_2O	lithium oxide
barium sulfide	BaS	$AlCl_3$	aluminum chloride
lithium fluoride	LiF	Mg_3P_2	magnesium phosphide
manganese(III) ion	Mn^{3+}	Cu^{2+}	copper(II) ion
phosphorus tribromide	PBr_3	Na_2S	sodium sulfide
perchloric acid	$HClO_4$	S_2F_2	disulfur difluoride
ammonium phosphate	$(NH_4)_3PO_4$	H_2S	hydrosulfuric acid
iron(II) nitrate	$Fe(NO_3)_2$	HIO_3	iodic acid
potassium bromite	$KBrO_2$	Na_2TeO_3	sodium tellurite
sodium hydrogen carbonate	$NaHCO_3$	$Mg(H_2PO_4)_2$	magnesium dihydrogen phosphate
sulfuric acid	H_2SO_4	$PbCl_2$	lead(II) chloride
hydrofluoric acid	HF	K_2CO_3	potassium carbonate
potassium iodide	KI	Si_2F_6	disilicon hexafluoride
oxygen difluoride	OF_2	CuI	copper(I) iodide
magnesium sulfate 7-hydrate	$MgSO_4 \cdot 7 H_2O$	$BaCl_2 \cdot 2 H_2O$	barium chloride 2-hydrate

Chapter 7

1) There are 2 nitrogen atoms and 5 oxygen atoms in one N_2O_5 molecule.
2) The formula for phosphorus pentafluoride is PF_5.
3) The answer is d. 4) The answer is a.
5) KCl: 39.1 + 35.5 = 74.6 g/mol

6) 1.06×10^{24} molecules CO $\times \dfrac{1 \text{ mol CO}}{6.02 \times 10^{23} \text{ molecules CO}} \times \dfrac{28.0 \text{ g CO}}{1 \text{ mol CO}} = 49.3$ g CO

7) 1.62 g N $\times \dfrac{85.0 \text{ g NaNO}_3}{1 \text{ mol NaNO}_3} \times \dfrac{1 \text{ mol NaNO}_3}{1 \text{ mol N}} \times \dfrac{1 \text{ mol N}}{14.0 \text{ g N}} = 9.84$ g NaNO$_3$

8) 37.5 g K$_2$SO$_4$ $\times \dfrac{1 \text{ mol K}_2\text{SO}_4}{174.3 \text{ g K}_2\text{SO}_4} = 0.215$ mol K$_2$SO$_4$

9) Ba(OH)$_2$

Element	Grams		Percent	
Ba	$1 \times 137.3 = 137.3$	g Ba	$\dfrac{137}{171.3} \times 100 =$	80.2 % Ba
O	$2 \times 16.0 = 32.0$	g O	$\dfrac{32.0}{171.3} \times 100 =$	18.7 % O
H	$2 \times 1.01 = \underline{2.02}$	g H	$\dfrac{2.02}{171.3} \times 100 =$	$\underline{1.2 \text{ \% H}}$
	171.3	g Ba(OH)$_2$		100.1 %

10) Both CH$_3$ and C$_{30}$H$_{50}$O could be simplest formulas; the others are not simplest, or empirical, formulas because they are not in "lowest terms."

11)

Element	Grams	Moles	Mole Ratio	Formula Ratio	Simplest Formula	Molecular Formula
C	92.25	$\dfrac{92.25}{12.0} = 7.69$	1.00	1		
H	7.75	$\dfrac{7.75}{1.01} = 7.67$	1.00	1	CH	C$_6$H$_6$

Chapter 8

1) 2 Ca(s) + O$_2$(g) → 2 CaO(s)
2) Ba(OH)$_2$(s) → BaO(s) + H$_2$O(ℓ)
3) C$_4$H$_9$CHO(g) + 7 O$_2$(g) → 5 CO$_2$(g) + 5 H$_2$O(g)
4) 2 Li(s) + 2 H$_2$O(ℓ) → H$_2$(g) + 2 LiOH(aq)
5) 2 KOH(aq) + Cu(NO$_3$)$_2$(aq) → Cu(OH)$_2$(s) + 2 KNO$_3$(aq)
6) H$_2$SO$_4$(aq) + 2 NaOH(aq) → Na$_2$SO$_4$(aq) + 2 HOH(ℓ)
7) 2 HCl(aq) + CaCO$_3$(s) → CaCl$_2$(aq) + CO$_2$(g) + H$_2$O(ℓ)

Chapter 9

1) GIVEN: 3 mol C$_5$H$_{10}$ WANTED: mol CO$_2$ PATH: mol C$_5$H$_{10}$ → mol CO$_2$
FACTOR: 10 mol CO$_2$/ 2 mol C$_5$H$_{10}$ 3 mol C$_5$H$_{10}$ $\times \dfrac{10 \text{ mol CO}_2}{2 \text{ mol C}_5\text{H}_{10}} = 15$ mol CO$_2$

2) GIVEN: 12.0 mol CO_2 WANTED: g O_2 PATH: mol $CO_2 \to$ mol $O_2 \to$ g O_2
FACTORS: 15 mol O_2/ 10 mol CO_2; 32.0 g/mol O_2

$$12.0 \text{ mol } CO_2 \times \frac{15 \text{ mol } O_2}{10 \text{ mol } CO_2} \times \frac{32.0 \text{ g } O_2}{1 \text{ mol } O_2} = 576 \text{ g } O_2$$

3) GIVEN: 4.12 g O_2 WANTED: g H_2O PATH: g $O_2 \to$ mol $O_2 \to$ mol $H_2O \to$ g H_2O
FACTORS: 32.0 g/mol O_2; 10 mol H_2O/ 15 mol O_2; 18.0 g/mol H_2O

$$4.12 \text{ g } O_2 \times \frac{1 \text{ mol } O_2}{32.0 \text{ g } O_2} \times \frac{10 \text{ mol } H_2O}{15 \text{ mol } O_2} \times \frac{18.0 \text{ g } H_2O}{1 \text{ mol } H_2O} = 1.55 \text{ g } H_2O$$

4) GIVEN: 6.81 g C_5H_{10} WANTED: g CO_2
PATH: g $C_5H_{10} \to$ mol $C_5H_{10} \to$ mol $CO_2 \to$ g CO_2
FACTORS: 70.1 g/mol C_5H_{10}; 10 mol CO_2/ 2 mol C_5H_{10}; 44.0 g/mol CO_2

$$6.81 \text{ g } C_5H_{10} \times \frac{1 \text{ mol } C_5H_{10}}{70.1 \text{ g } C_5H_{10}} \times \frac{10 \text{ mol } CO_2}{2 \text{ mol } C_5H_{10}} \times \frac{44.0 \text{ g } CO_2}{1 \text{ mol } CO_2} = 21.4 \text{ g } CO_2$$

GIVEN: 19.2 g (act); 21.4 g (theo)   WANTED: % yield

EQUATION: % yield = $\frac{\text{actual yield}}{\text{theoretical yield}} \times 100 = \frac{19.2 \text{ g}}{21.4 \text{ g}} \times 100 = 89.7\%$

5)
	4 Fe	+	3 O_2	→	2 Fe_2O_3
Grams at start	15.2		7.41		0
Molar Mass (g/mole)	55.9		32.0		159.8
Moles at start	0.272		0.232		0
Moles used(-), produced(+)	-0.272		-0.204		+0.136
Moles at end	0.000		0.028		0.136
Grams at end (See below.)	0.000		0.90		21.7

The Fe is the limiting reagent, and 21.7 g Fe_2O_3 can be produced.

6) The O_2 is in excess. The mass of O_2 remaining after reaction is 0.90 g.

7) GIVEN: 127 J   WANTED: kcal   PATH: J $\to$ cal $\to$ kcal
FACTORS: 4.184 J/cal; 1000 cal/kcal

$$127 \text{ J} \times \frac{1 \text{ cal}}{4.184 \text{ J}} \times \frac{1 \text{ kcal}}{1000 \text{ cal}} = 0.0304 \text{ kcal}$$

8) $H_2O(g) \to H_2(g) + 1/2\ O_2(g)$ $\Delta H = +286$ kJ;   286 kJ + $H_2O(g) \to H_2(g) + 1/2\ O_2(g)$

9) GIVEN: 2.22 kJ   WANTED: g C_5H_{10} PATH: kJ $\to$ mol $C_5H_{10} \to$ g C_5H_{10}
FACTORS: 6.08 kJ/ 2 mol C_5H_{10}; 70.1 g/mol C_5H_{10}

$$2.22 \text{ kJ} \times \frac{2 \text{ mol } C_5H_{10}}{6.08 \text{ kJ}} \times \frac{70.1 \text{ g } C_5H_{10}}{1 \text{ mol } C_5H_{10}} = 51.2 \text{ g } C_5H_{10}$$

Chapter 10

1) The answer is b.   2) The answer is a.   3) The answer is d.   4) The answer is b.

5) The alkali metals share an ns^1 highest occupied energy level electron configuration. In this description, n is the principal quantum number of the highest occupied energy level when the atom is in its ground state. For lithium, $n = 2$; for sodium, $n = 3$; for potassium, $n = 4$, and so forth. The s refers to the s sublevel in the highest occupied energy level. The p, when needed, refers to the p sublevel in the same energy level. Superscript numbers identify the number of electrons in those sublevels. The ns^1 indicates that the highest occupied energy level of a ground state alkali metal is occupied by one electron in the s orbital.

6) The Lewis symbol for oxygen is :Ö:

Any of the following are Lewis symbols of an alkaline earth element: Be: Mg: Ca: Sr: Ba: Ra:

7) Cl: $1s^2 2s^2 2p^6 3s^2 3p^5$ or $[Ne]3s^2 3p^5$

8) Ti: $1s^2 2s^2 2p^6 3s^2 3p^6 4s^2 3d^2$ or $[Ar]4s^2 3d^2$ The 3d may precede the 4s.

9) The answer is d.   10) The answer is d.   11) The ions S^{2-}, Cl^- and Ca^{2+} are arranged in order of decreasing size, largest to smallest.

12) The answer is d.   13) The answer is a.

14) The metals have Z = 11, 20 and 43; the non-metals have Z = 7, 16 and 53.

Chapter 11

1) The answer is e.   2) The answer is a.   3) The answer is c.
4) The answer is d.   5) The answer is b.   6) The answer is a.

Chapter 12

1) :F̈—Ö:
     |
     :F̈:

2) H—N̈—H
       |
       H

3) $\begin{bmatrix} :\ddot{O}: \\ | \\ :\ddot{O}—Se—\ddot{O}: \\ | \\ :\ddot{O}: \end{bmatrix}^{2-}$

4) $\begin{bmatrix} \ddot{O}=N—\ddot{O}: \\ | \\ :\ddot{O}: \end{bmatrix}^{-}$

5) H—Ö—P—Ö—H
         |
         :Ö—H

6) H—C≡C—C—H with H's on the last C (top, bottom, side)

   OR   H₂C=C=CH₂ form

7) H—C—C—C—H with H's and :F̈: substituents   OR   H—C—C—C—F̈:

8) H—C—C—C—Ö—H with =O on third C

	Electron Pair Geometry		Molecular Geometry
9)	tetrahedral	(OF_2)	bent
10)	tetrahedral	(NH_3)	pyramidal
11)	tetrahedral	(SeO_4^{2-})	tetrahedral
12)	trigonal planar	(NO_3^-)	trigonal planar

13)
```
        F δ−
        /
δ+ H—B δ+
        \
        F δ−
```
The BHF_2 molecule is polar from the arrangement of the B—F bonds.

14)
```
δ−   δ+   δ−
Br—Be—Br
```
The $BeBr_2$ molecule is nonpolar; the polar bonds cancel each other out.

15)
```
  H   H H H              H H H H                H H
  |   | | |              | | | |                | |
  C=C—C—C—H          H—C—C=C—C—H            H—C—C—H
  |     | |              |     |                | |
  H     H H              H     H            H—C—C—H
                                                | |
                                                H H
```
```
    H H
    | |
H—C—C—H
    \ /
     C
    / \
   H   CH₃
```

The top left and top center molecules are alkenes; note the carbon-carbon double bonds. The molecules top right and bottom left are cycloalkanes; note the carbon atoms form a closed cycle, or ring.

16)
```
    H H                    H     H
    | |                    |     |
H—C—C—Ö—H            H—C—Ö—C—H
    | |                    |     |
    H H                    H     H
```

The molecule at the left is an alcohol; the right hand molecule is an ether.

Chapter 13

1) The answer is b. According to Avogadro's Law, equal volumes of gases at the same temperature and pressure contain an equal number of molecules (choice a) or moles of molecules (choice c). Both oxygen and chlorine are diatomic, so the number of chlorine and oxygen atoms is equal (choice d). Because the masses of oxygen and chlorine molecules are different, the masses of equal numbers of molecules are also different.

2) $PV = nRT$; $PV/nT = R$

3) GIVEN: 0°C (273 K); 760 torr; 17.0 g/mol NH_3; 62.4 L · torr/mol · K
   WANTED: Density in g/L (m/V)

Equation: $\dfrac{m}{V} = \dfrac{(MM)\,P}{RT} = \dfrac{17.0\text{ g}}{\text{mol}} \times \dfrac{760\text{ torr}}{273\text{ K}} \times \dfrac{\text{mol}\cdot\text{K}}{62.4\text{ L}\cdot\text{torr}} = 0.758\text{ g/L}$

4) GIVEN: 0.369 g; 0.460 L; 22°C (295 K); 819 torr; 62.4 L·torr/mol·K
WANTED: molar mass, MM, units of g/mol

EQUATION: $MM = \dfrac{mRT}{PV} = \dfrac{0.369\text{ g}}{819\text{ torr}} \times \dfrac{62.4\text{ L}\cdot\text{torr}}{\text{mol}\cdot\text{K}} \times \dfrac{295\text{ K}}{0.460\text{ L}} = 18.0\text{ g/mol}$

5) GIVEN: −17°C (256 K); 1.03 atm; 0.0821 L·atm/mol·K   WANTED: Molar Volume, V/n

EQUATION: Molar Volume (MV) $= \dfrac{V}{n} = \dfrac{RT}{P} = \dfrac{0.0821\text{ L}\cdot\text{atm}}{\text{mol}\cdot\text{K}} \times \dfrac{256\text{ K}}{1.03\text{ atm}} = 20.4\text{ L/mol}$

6) GIVEN: 1.25 L   WANTED: mol   PATH: mol/L → mol   FACTOR: 23.2 L/mol

$1.25\text{ L} \times \dfrac{1\text{ mol}}{23.2\text{ L}} = 0.0539\text{ mol}$

7) GIVEN: 767 torr; 24°C (297 K); 62.4 L·torr/mol·K   WANTED: MV, V/n, units of L/mol

EQUATION: $MV = \dfrac{V}{n} = \dfrac{RT}{P} = \dfrac{62.4\text{ L}\cdot\text{torr}}{\text{mol}\cdot\text{K}} \times \dfrac{297\text{ K}}{767\text{ torr}} = 24.2\text{ L/mol}$

GIVEN: 0.148 L H_2 WANTED: g Zn   PATH: L H_2 → mol H_2 → mol Zn → g Zn
FACTORS: 24.2 L H_2/mol; 1 mol Zn/ 1 mol H_2; 65.4 g Zn/mol Zn

$0.148\text{ L }H_2 \times \dfrac{1\text{ mol }H_2}{24.2\text{ L }H_2} \times \dfrac{1\text{ mol Zn}}{1\text{ mol }H_2} \times \dfrac{65.4\text{ g Zn}}{1\text{ mol Zn}} = 0.400\text{ g Zn}$

OR... Using the Second Method...

GIVEN: 0.148 L H_2; 767 torr; 24°C (297 K); 62.4 L·torr/mol·K   WANTED: mol H_2

EQUATION: $n = \dfrac{PV}{RT} = \dfrac{767\text{ torr} \times 0.148\text{ L}}{(62.4\text{ L}\cdot\text{torr/mol}\cdot\text{K}) \times 297\text{ K}} = 0.00613\text{ mol }H_2$

GIVEN: 0.00613 mol H_2 WANTED: g Zn   PATH: mol H_2 → mol Zn → g Zn
FACTORS: 1 mol Zn/ 1 mol H_2; 65.4 g/mol Zn

$0.00613\text{ mol }H_2 \times \dfrac{1\text{ mol Zn}}{1\text{ mol }H_2} \times \dfrac{65.4\text{ g Zn}}{1\text{ mol Zn}} = 0.401\text{ g Zn}$

8)

	Volume	Temperature	Pressure	Amount
Initial Value (1)	4.18 L	38°C = 358 K	1.34 atm	constant
Final Value (2)	V_2	18°C = 301 K	0.891 atm	constant

$4.18\text{ L }C_4H_{10} \times \dfrac{(273 + 18)\text{ K}}{(273 + 38)\text{ K}} \times \dfrac{1.34\text{ atm}}{0.891\text{ atm}} \times \dfrac{13\text{ L }O_2}{2\text{ L }C_4H_{10}} = 38.2\text{ L }O_2$

9) P = 241 + 336 + 413 = 991 torr   10) 2174 = 151 + 1162 + p_{O_2} ; p_{O_2} = 861 torr

Chapter 14

1) The answer is b.   2) The answer is c.   3) The answer is b.   4) The answer is b.
5) The answer is a.   6) The answer is d.   7) The answer is a.   8) The answer is a.
9) The answer is c.   10) The answer is d.

11) GIVEN: 10.0 kJ; 2.35 g Na   WANTED: ΔH_{vap} in kJ/g

    EQUATION: $Q = m\Delta H$, $\Delta H = \dfrac{Q}{m} = \dfrac{10.0 \text{ kJ}}{2.35 \text{ g}} = 4.26$ kJ/g

12) GIVEN: 123 g Ni; $\Delta H_{fus} = 310$ J/g   WANTED: Q in kJ
    EQUATION: $Q = m\Delta H_{fus} = (123 \text{ g})(310 \text{ J/g}) = 38{,}130$ J $= 38.1$ kJ

13) Liquid only exists in region 3–4; the $\Delta H_{vaporization}$ is region 4–5.

14) GIVEN: 3636 g; 3.6 J/g·°C; $T_i = 16°C$, $T_f = 110°C$ WANTED: Q
    EQUATION: $Q = mc\Delta T = (3636 \text{ g})(3.6 \text{ J/g}\cdot°C) = 1.2 \times 10^6$ J $= 1.2 \times 10^3$ kJ

15) GIVEN: 65.0 g; -1.92 kJ; $T_i = 114.6°C$, $T_f = 31.2°C$ WANTED: c (J/g·°C)

    EQUATION: $c = \dfrac{Q}{m\Delta T} = \dfrac{-1920 \text{ J}}{(65.0 \text{ g})(-83.4°C)} = 0.354$ J/g·°C

16) GIVEN: 44 g $H_2O(\ell)$, $T_i = 82°C$, $T_f = -23°C$, freezing point = 0°C
    Specific heats: 2.1 J/g·°C (s), 4.18 J/g·°C, (ℓ), $\Delta H_{fus} = 335$ J/g
    WANTED: total heat flow Q_{total} EQUATIONS: $Q = mc\Delta T$, $Q = m\Delta H_{fus}$

$Q_{total} = Q_{water} + Q_{\Delta fus} + Q_{ice}$
$Q_{water} = (44 \text{ g})(4.18 \text{ J/g·°C})(-82°C)$ = -15,081 J = -15. kJ
$Q_{\Delta fus} = (44 \text{ g})(-335 \text{ J/g})$ = -14,740 J = -15. kJ
$Q_{ice} = (44 \text{ g})(2.1 \text{ J/g·°C})(-23°C)$ = -2,125 J = <u>- 2.1 kJ</u>
                                                          Q_{total} = -32. kJ

Chapter 15

1) The answer is c.   2) The answer is a.   3) The answer is c.   4) The answer is a.
5) The answer is c.

6) GIVEN: 312 g solution   WANTED: g NaCl, g H_2O PATH: g solution → g NaCl
    FACTOR: 0.90 g NaCl/100 g solution   g NaCl = 312 g soln × $\dfrac{0.90 \text{ g NaCl}}{100 \text{ g soln}} = 2.8$ g

    312 g soln = g H_2O + g NaCl; 312 g = g H_2O + 3 g   g H_2O = 312 - 3 = 309 g

7) GIVEN: 0.716 mol H_2SO_4 WANTED: mL H_2SO_4 PATH: mol → L → mL

    FACTOR: 0.415 mol H_2SO_4/L   0.716 mol H_2SO_4 × $\dfrac{1 \text{ L } H_2SO_4}{0.415 \text{ mol } H_2SO_4}$ × $\dfrac{1000 \text{ mL}}{1 \text{ L}} = 1730$ mL

8) GIVEN: M_c = 12.0 mol/L_c; V_c = 0.0625 L_c; M_d = 0.812 mol/L_d WANTED: V_d

EQUATION: $V_d = \dfrac{V_c \times M_c}{M_d} = \dfrac{0.0625\ L_c \times 12.0\ \text{mol}/L_c}{0.812\ \text{mol}/L_d} = 0.924\ L = 924\ mL$

9) GIVEN: 24.16 mL (0.02416 L) Na_2CO_3 WANTED: L CO_2
PATH: L $Na_2CO_3 \to$ mol $Na_2CO_3 \to$ mol $CO_2 \to$ L CO_2
FACTORS: 0.0872 mol Na_2CO_3/L; 1 mol CO_2/mol Na_2CO_3

$0.02416\ L\ Na_2CO_3 \times \dfrac{0.0872\ \text{mol}\ Na_2CO_3}{1\ L\ Na_2CO_3} \times \dfrac{1\ \text{mol}\ CO_2}{1\ \text{mol}\ Na_2CO_3} = 0.00211\ \text{mol}\ CO_2$

GIVEN: 0.00211 mol CO_2; 756 torr; 23°C (296 K); 62.4 L·torr/mol·K   WANTED: L CO_2

EQUATION: $V = \dfrac{nRT}{P} = \dfrac{0.00211\ \text{mol} \times 62.4\ L\cdot\text{torr/mol}\cdot K \times 296\ K}{756\ \text{torr}} = 0.0515\ L = 51.5\ mL\ CO_2(g)$

This solution uses the "Second Method" on text pp. 346–347.

10) GIVEN: 14.37 mL (0.01437 L) Na_2CO_3 WANTED: L HCl
PATH: L $Na_2CO_3 \to$ mol $Na_2CO_3 \to$ mol HCl $\to$ L HCl
FACTORS: 0.102 mol Na_2CO_3/L; 0.123 mol HCl/L; 2 mol HCl/mol Na_2CO_3

$0.01437\ L\ Na_2CO_3 \times \dfrac{0.102\ \text{mol}\ Na_2CO_3}{1\ L\ Na_2CO_3} \times \dfrac{2\ \text{mol}\ HCl}{1\ \text{mol}\ Na_2CO_3} \times \dfrac{1\ L\ HCl}{0.123\ \text{mol}\ HCl} = 0.0238\ L\ HCl$

11) GIVEN: 0.317 g Na_2CO_3; 19.46 mL (0.1946 L) HCl   WANTED: M = mol HCl/L HCl
PATH: g $Na_2CO_3 \to$ mol $Na_2CO_3 \to$ mol HCl $\to$ mol HCl/L
FACTORS: 106 g/mol Na_2CO_3; 2 mol HCl/mol Na_2CO_3

$0.317\ g\ Na_2CO_3 \times \dfrac{1\ \text{mol}\ Na_2CO_3}{106\ g\ Na_2CO_3} \times \dfrac{2\ \text{mol}\ HCl}{1\ \text{mol}\ Na_2CO_3} \times \dfrac{1}{0.01946\ L\ HCl} = 0.307\ M\ HCl$

12) GIVEN: 11.7 mL (0.0117 L) Na_2CO_3; 22.4 mL (0.0224 L) HCl   WANTED: M = mol HCl/L HCl
PATH: L $Na_2CO_3 \to$ mol $Na_2CO_3 \to$ mol HCl $\to M_{HCl}$ = mol HCl/L HCl
FACTORS: 0.113 mol Na_2CO_3/L Na_2CO_3; 2 mol HCl/1 mol Na_2CO_3

$0.0117\ L\ Na_2CO_3 \times \dfrac{0.113\ \text{mol}\ Na_2CO_3}{1\ L\ Na_2CO_3} \times \dfrac{2\ \text{mol}\ HCl}{1\ \text{mol}\ Na_2CO_3} \times \dfrac{1}{0.0224\ L\ HCl} = 0.118\ M\ HCl$

Molality and Colligative Properties

1) GIVEN: 1.60 g $HC_2H_3O_2$; 21.47 g (0.02147 kg) solvent; ΔT_b = 3.14°C   WANTED: K_b
PATH: g $HC_2H_3O_2 \to$ mol $HC_2H_3O_2$ FACTOR: 60.0 g/mol $HC_2H_3O_2$
EQUATIONS: m = mol $HC_2H_3O_2$/kg solvent; $\Delta T_b = K_b m$; $K_b = \Delta T_b/m$

$m = \dfrac{1.60\ g\ HC_2H_3O_2 / (60.0\ g\ HC_2H_3O_2/\text{mol})}{0.02147\ kg} = 1.24\ m$; $\dfrac{3.14°C}{1.24} = 2.53\ °C/m = K_b$

2) GIVEN: 5.22 g $C_{10}H_8$; 128 g $C_{10}H_8$/mol; 35.81 g (0.03581 kg) solvent
WANTED: m (mol $C_{10}H_8$/kg solvent)   PATH: g $C_{10}H_8 \to$ mol $C_{10}H_8 \to$ mol/kg

$\dfrac{5.22\ g\ C_{10}H_8}{0.03581\ \text{kg solvent}} \times \dfrac{1\ \text{mol}\ C_{10}H_8}{128\ g\ C_{10}H_8} = 1.14\ \dfrac{\text{mol}\ C_{10}H_8}{\text{kg solvent}}$ so m = 1.14

3) GIVEN: m = 1.14; K_f = 20.2°C/m   WANTED: ΔT_f, then new freezing point

   EQUATION: $\Delta T_f = K_f m$ = (-20.2°C/m)(1.14m) = -23.0°C; new freezing pt. is (6.5 - 23.0) = -16.5°C

4) GIVEN: ΔT_f = 4.31°C; K_f = 5.48°C/m   WANTED: m, molality

   EQUATION: $m = \dfrac{\Delta T_f}{K_f} = \dfrac{4.31°C}{5.48°C/m}$ = 0.786 m = 0.786 mol solute/ kg solvent

   GIVEN: 0.786 m; 4.18 g solute; 19.89 g (0.01929 kg) solvent   WANTED: MM solute, g/mol

   $\dfrac{4.18 \text{ g solute}/ 0.01989 \text{ kg solvent}}{0.786 \text{ mol solute/kg solvent}} = \dfrac{267 \text{ g solute}}{\text{mol solute}}$ = 267 g/mol

Normality

1) GIVEN: 2 equivalents/mol Na_2CO_3 (from equation) WANTED: g Na_2CO_3/eq
   PATH: g/mol → g/eq   FACTOR: 106.0 g Na_2CO_3/mol Na_2CO_3

   $\dfrac{106 \text{ g } Na_2CO_3}{1 \text{ mol } Na_2CO_3} \times \dfrac{1 \text{ mol } Na_2CO_3}{2 \text{ eq}} = 53.0 \dfrac{\text{g } Na_2CO_3}{\text{eq}}$ = equivalent mass Na_2CO_3

2) GIVEN: 17.9 mL (0.0179 L) Na_2CO_3 WANTED: eq   PATH: L Na_2CO_3 → eq
   FACTOR: 0.119 eq/ L $0.0179 \text{ L } Na_2CO_3 \times \dfrac{0.119 \text{ eq}}{1 \text{ L } Na_2CO_3}$ = 0.00213 eq

3) GIVEN: 0.288 g $NaC_2H_3O_2$ (NaOAc); 21.42 mL (0.02142 L) HCl   WANTED: N HCl (eq/L)
   PATH: g $NaC_2H_3O_2$ → mol $NaC_2H_3O_2$ → eq $NaC_2H_3O_2$ → eq HCl → eq HCl/L
   FACTORS: 106 g $NaC_2H_3O_2$/mol; 2 eq/mol $NaC_2H_3O_2$

   $\dfrac{0.288 \text{ g NaOAc}}{0.02142 \text{ L HCl}} \times \dfrac{1 \text{ mol NaOAc}}{106 \text{ g NaOAc}} \times \dfrac{2 \text{ eq base}}{1 \text{ mol NaOAc}} \times \dfrac{1 \text{ eq acid}}{1 \text{ eq base}}$ = 0.254 N HCl

4) GIVEN: V_1 = 25.1 mL (0.0251 L_1), N_1 = 0.0995 eq/ L_1; V_2 = 11.3 mL (0.0113 L_2)   WANTED: N_2

   EQUATION: $V_2 N_2 = V_1 N_1$ $N_2 = \dfrac{V_1 N_1}{V_2} = \dfrac{(0.0251 L_1)(0.0995 \text{ eq}/ L_1)}{(0.0113 L_2)}$ = 0.221 N Na_2CO_3

Chapter 16

1) The answer is c.
2) $CaCl_2$: Ca^{2+}(aq) + 2 Cl^-(aq)   HI: H^+(aq) + I^-(aq)
   $HC_2H_3O_2$: $HC_2H_3O_2$(aq) $Al(NO_3)_3$: Al^{3+}(aq) + 3 NO_3^-(aq)
3) 2 Na^+(aq) + SO_4^{2-}(aq) + 2 K^+(aq) + CO_3^{2-}(aq) → NR
4) 2 H^+(aq) + CO_3^{2-}(aq) → CO_2(g) + H_2O(ℓ)
5) H^+(aq) + CHO_2^-(aq) → $HCHO_2$(aq)
6) Pb^{2+}(aq) + SO_4^{2-}(aq) → $PbSO_4$(s)
7) $Ca(OH)_2$(s) + 2 H^+(aq) → Ca^{2+}(aq) + 2 HOH(ℓ)
8) Ni(s) + 2 H^+(aq) → Ni^{2+}(aq) + H_2(g)

Chapter 17

1) The answer is b.
2) The answer is c.
3) The answer is a.
4) The answer is d.
5) The answer is d.
6) The answer is c.
7) The answer is a.
8) The answer is b.
9) The answer is b.
10) The answer is d.
11) The answer is a.

12) $HCN(aq) + SO_3^{2-}(aq) \rightleftarrows HSO_3^-(aq) + CN^-(aq)$ favored direction   LEFT

13) $HSO_3^-(aq) + HPO_4^{2-}(aq) \rightleftarrows SO_3^{2-}(aq) + H_2PO_4^-(aq)$ favored direction   LEFT

     $HSO_3^-(aq) + HPO_4^{2-}(aq) \rightleftarrows H_2SO_3(aq) + PO_4^{3-}(aq)$ favored direction   LEFT

14) pOH = 11.13   15) $[H^+] = 1.3 \times 10^{-3}$ 16) $[OH^-] = 7.4 \times 10^{-12}$

Chapter 18

1) $Al^{3+}(aq) + 3\ e^- \rightarrow Al(s)$ RED(UCTION)
      $Ca(s) \rightarrow Ca^{2+}(aq) + 2\ e^-$ OX(IDATION)

2)
$$3\ Ca(s) \rightarrow 3\ Ca^{2+}(aq) + 6\ e^-$$
$$2\ Al^{3+}(aq) + 6\ e^- \rightarrow 2\ Al(s)$$
$$\overline{3\ Ca(s) + 2\ Al^{3+}(aq) \rightarrow 3\ Ca^{2+}(aq) + 2\ Al(s)}$$

3) Cu, 0, 2+
4) Cr, 6+, 3+
5) CrO_4^{2-}, Cu
6) The answer is b.

7a) As oxidizers, $X^+ > Y^+$ 7b) as reducers, $Y > W^-$ 8) Answers are b and c.

9) (a) $6\ I^-(aq) \rightarrow 3\ I_2(aq) + 6\ e^-$
   (b) $6\ H^+(aq) + ClO_3^-(aq) + 6\ e^- \rightarrow Cl^-(aq) + 3\ H_2O(\ell)$
   (c) $\overline{6\ I^-(aq) + 6\ H^+(aq) + ClO_3^-(aq) \rightarrow 3\ I_2(aq) + Cl^-(aq) + 3\ H_2O(\ell)}$

Chapter 19

1) The answer is b.
2) The answer is d.
3) The answer is c.
4) The answer is a.
5) The answer is b.
6) The answer is a.
7) The answer is b.
8) The answer is c.

9) $K = \dfrac{[CO_2]^2}{[CO]^2[O_2]}$ 10) $K = [Pb^{2+}][Br^-]^2$ 11) $K = \dfrac{[NH_4^+][OH^-]}{[NH_3]}$

12) For silver sulfate, $Ag_2SO_4(s) \rightleftarrows 2\ Ag^+(aq) + SO_4^{2-}(aq)$, $K_{sp} = 1.2 \times 10^{-5}$.
Let s = solubility of Ag_2SO_4, in moles per liter, then $[Ag^+] = 2s$, $[SO_4^{2-}] = s$
$K_{sp} = 1.2 \times 10^{-5} = [Ag^+]^2[SO_4^{2-}] = (2s)^2(s)$
      $1.2 \times 10^{-5} = 4s^3$
        $1.4 \times 10^{-2} = s$

13) $[H^+] = (K_a[HCHO_2])^{1/2}$
    $= (2.0 \times 10^{-4} \cdot 1.2)^{1/2}$
    $= (2.4 \times 10^{-4})^{1/2}$
    $= 1.5 \times 10^{-2}$
    pH = -log $[H^+]$ = -log (1.5×10^{-2}) = 1.81

14) 6.10 g $HCHO_2 \times \dfrac{1 \text{ mol } HCHO_2}{46.0 \text{ g } HCHO_2} = \dfrac{0.133 \text{ mol}}{HCHO_2}$ 43.1 g $NaCHO_2 \times \dfrac{1 \text{ mol } NaCHO_2}{68.0 \text{ g } NaCHO_2} = \dfrac{0.634 \text{ mol}}{NaCHO_2}$

Let V = solution volume, in L, then $[HCHO_2]$ = 0.133/V, and $[CHO_2^-]$ = 0.634/V, then using the K_a expression and K_a value from Question 13, we obtain

$\dfrac{[H^+](0.634/V)}{(0.133/V)} = 2.0 \times 10^{-4}$ $[H^+] = \dfrac{2.0 \times 10^{-4}(0.133)}{(0.634)}$

$[H^+] = 4.2 \times 10^{-5}$
pH = 4.38

15) $K = \dfrac{[NH_3]^2}{[N_2][H_2]^3} = \dfrac{(0.042)^2}{(0.057)(0.17)^3} = \dfrac{(0.00176)}{(0.057)(0.00491)} = 6.3$

Chapter 20

1) The answer is c.   2) The answer is d.   3) The answer is a.   4) The answer is a.

5) GIVEN: time 24 hours; R = 0.20 mg, S = 3.7 mg   WANTED: $t_{1/2}$ (hours/half-life)

$\dfrac{R}{S} = \dfrac{0.20}{3.7} = 0.054$ of the sample remains   From Figure 20.4, this is about 4.2 or 4.3 half-lives.

$\dfrac{24 \text{ hrs}}{4.2 \text{ half-lives}} = 5.7$ hrs/half-life

6) The answer is b.   7) The answer is c.   8) The answer is c.   9) The answer is c.
10) The answer is d.   11) The answer is b.   12) The answer is a.   13) The answer is a.
14) The answer is c.

Chapter 21

1) The answer is d.   2) The answer is c.   3) The answer is d.
4) 1,1-diethyl-3-methylcyclohexane is below.   5) The answer is b.   6) the answer is b.
                                                  7) The answer is c.   8) Mesitylene is below.

9) CH$_3$-CH$_2$-CH$_2$-CH$_2$-CH$_2$-CH$_2$-OH 1-hexanol   6-hexanol is incorrect for this isomer.
    CH$_3$-CH$_2$-CH$_2$-CH$_2$-CH-CH$_3$ 2-hexanol   5-hexanol is incorrect for this isomer.
                              |
                              OH
    CH$_3$-CH$_2$-CH$_2$-CH-CH$_2$-CH$_3$ 3-hexanol   4-hexanol is incorrect for this isomer.
                        |
                        OH

10) The answer is d.   11) The answer is a.   12) The answer is d.   13) The answer is c.

14) $\left[\begin{array}{cccccc} F & F & F & F & F & F \\ | & | & | & | & | & | \\ -C-&C-&C-&C-&C-&C- \\ | & | & | & | & | & | \\ Cl & F & Cl & F & Cl & F \end{array} \right]_n$

15) Structure: CH=CH attached to N of pyrrolidinone ring (N-vinyl-2-pyrrolidone)

Chapter 22

1) The answer is b.   2) The answer is d.   3) The answer is b.   4) The answer is d.
5) The answer is c.   6) The answer is d.   7) The answer is e.   8) The answer is b.
9) The answer is a.   10) The answer is c.   11) The answer is d.   12) The answer is b.
13) The answer is c.   14) The answer is a.